THE BAOFENG RADIO BIBLE

The Comprehensive and Easy-to-Follow Guerrilla's Guide to Becoming a Pro with Your Baofeng Radio in No Time and Staying Connected When It Matters Most

Eliot J. Hawke

TABLE OF CONTENTS

Hey there!

Big thanks for picking up my book on Baofeng Radios! I am excited that you are getting into the world of radio communication, and I've got a little something extra for you.

Free Bonuses Just a QR Code Away!

I want to make sure you get the most out of your Baofeng Radio, so I've put together some awesome freebies

- **Emergency Communication Plan Template** - Stay prepared for anything
- **Maintenance Checklist** - Keep your radio in tip-top shape
- **Step-by-Step Programming Tutorials** - Programming made easy
- **Troubleshooting Guide** - Quick fixes at your fingertips

Grab Your Bonuses

Just scan the QR code right here on this page with your phone, and you're all set. It's a small thank you from me, Eliot Hawke, for joining the Baofeng community.

Enjoy the extras and happy communicating!

Cheers,

Eliot

Introduction

Welcome to the essential guide for mastering the Baofeng radio— a technological masterpiece that is not only budget-friendly but also your key to seamless and powerful communication in all crucial situations.

The Baofeng radio is much more than your ordinary gadget. Equipped with the ability to operate from Very High Frequency (VHF) to Ultra-High Frequency (UHF), from Frequency Modulation (FM) to analog communication, this tool is a powerhouse that guarantees you will be able to open the gates to a world of tactical communication possibilities.

Through a methodical structure, this book will guide you across the path to becoming a radio pro with minimal effort required. Within these pages, you will find all the details you need to understand the strengths of the Baofeng while also discovering how it can elevate your inner ability to communicate effectively in high-pressure situations and critical scenarios.

However, in order to understand the extent of the power that you would obtain by mastering the Boafeng radio, it is essential to become aware of the fact that communication is the lifeline in any mission or survival situation. On this note, we can recognize three crucial types of communication—sustainment, tactical, and clandestine/strategic—each one of them playing a distinct role.

- *Sustainment communication* focuses on maintaining ongoing logistical support and coordination. An example of sustainment communication could involve a scenario where a team of explorers uses a radio to share information with base camp about supplies, weather forecasts, and the coordination of their route.

- *Tactical communication* concerns immediate communication during operations. Picture a SWAT team using the Baofeng radio during a high-risk search and arrest mission. Since the team would rely on instant updates, location sharing, and coordination of tactical movements to ensure effective response to changing situations, this is considered an instance of tactical communication.

- *Clandestine or strategic communication* is all about discreetly sharing information, which makes it essential for undercover operations or strategic planning. An example of strategic communication could involve the CIA using a radio to transmit encrypted messages securely, guaranteeing operational safety and confidentiality.

The Boafeng radio's versatility makes it the perfect tool to ensure flawless communication, regardless of the urgency and secrecy of the situation. Besides being portable, this device is packed with features that let you tailor your communication experience to each particular scenario, from exploring and storing multiple frequencies to engaging satellite communications and more.

In these pages, you will learn how to never feel isolated again. Even in environments where traditional networks might falter, you will be equipped with the knowledge necessary to remain in touch 100% of the time. Think of this book as your survival bible, providing you not only with bravery and resilience but also with the vital art of all-around communication.

Due to the many ways and scenarios in which the Baofeng radio can be utilized, mastering it might sound daunting at first. For this reason, this book is structured in a way that will make the learning process easy through concise explanations and concrete examples, ensuring you will not get discouraged while trying to become a radio master. Trust me, communicating with the Boafeng radio is a skill worth conquering, especially when your survival depends on your ability to remain connected.

The Baofeng radio is more than just features and wiring; it is about empowerment. Surely, as you delve deeper, you will earn the power to fully own the communication realm while you learn all you need to know about programming frequencies, crafting makeshift antennas, encrypting messages, and cutting through noise to transmit critical information.

By putting your full trust in these pages, radio frequencies will become your second language. You will be able to learn more about the different types of frequencies on which the Boafeng radio operates and how the radio's ability to communicate in VHF, UHF, FM, and analog communication makes it such an indispensable tool during all kinds of adventures. Now, let's quickly break down these key elements of radio communication to begin grasping the idea of how powerful this tool truly is:

HF (Very High Frequency) and UHF (Ultra-High Frequency) are specific bands within the electromagnetic spectrum. Baofeng radio operates on VHF (30-300 MHz) when the situation calls for long-distance communication, particularly in outdoor environment where signals need to travel further. Meanwhile, UHF range (300 MHz to 3 GHz) is mainly effective within urban areas thanks to its ability to penetrate structures. For this reason, Baofeng radio operates on UHF when there is the need to guarantee clear communication in crowded areas.

FM (Frequency Modulation) is a method used for transmitting information, particularly audio signals, via radio waves. This modulation technique is used to ensure top-quality transmission and is generally the leasing technique in broadcasting and two-way radio systems. Thanks to its FM, the Baofeng radio is a perfect tool to share clear information and announcements during public gatherings.

Analog communication, in contrast to digital methods, involves continuous signal modulation without digital processing. This is the most traditional form of communication, which relies on changing the amplitude or frequency of the carrier wave to carry information. Although the analog system is not considered advanced technology, the aofeng radio makes sure to support it in order to be compatible with traditional radio systems. This feature guarantees ccessful communication with other existing devices.

nderstanding these fundamental concepts is crucial for maximizing the capabilities of devices like the Baofeng radio, hich can use VHF, UHF, FM, and analog modes. By harnessing the unique properties of each frequency band and odulation technique, you can optimize your communication strategy for different environments and scenarios.

the upcoming chapters, you will learn all about how the Baofeng radio leverages these technologies to ensure reliable d effective communication in diverse settings. But beware that this book is more than just a simple guide; it represents chance for you to become part of a community of like-minded individuals who value technology, preparedness, and e timeless importance of human connection.

a world where conventional methods can fail, this book equips you with the tools to ensure sustainable ommunication, regardless of your skill level or the state of modern infrastructure.

re you ready to become a modern survivalist, always connected and prepared?

/ithout further ado, let's begin your adventure with the Baofeng radio!

Part I
Baofeng Radio Basics

CHAPTER 1
Origins of Baofeng Radios

Historical Evolution of Baofeng

Before we dive into how exactly you can use the Baofeng radio to communicate effectively, let's look at its history first. The origins of the Baofeng radio, or its manufacturing company, can be traced back to 1993. In that year, a company by the name Fujian Nan'an Baofeng Electronics Co., Ltd., laid its foundation. The company developed wireless communication equipment, but it wasn't until the year 2000 that they would prevail as one that redefined audio communication.

During that time a group of engineers led by Chief Executive Officer (CEO) Zheng Guangming, began experimenting with two-ways radio technology. A year later the company launched the UV-3R as its inaugural product and marked their entry into the consumer electronics market. The UV-3R was a product that was far from ordinary and truly one of a kind at the time.

What made the UV-3R so unique was that it pioneered the concept of dual-band functionality. Another great thing that made the product truly standout among others was its price. Along with being technologically sophisticated, the product was one that radio enthusiasts could easily afford. Having a product in the market that catered to both the price and functionality preferences is what marked the company's rise.

As the UV-3R gained market and consumer approval, Baofeng continued to innovate and excel beyond the limitations of conventional design protocols. With such innovation, the company began catering to the diverse needs of multiple market segments and target audiences. Common examples of segment and audiences included operators, emergency responders, and outdoor adventurers.

The company continued to innovate and as the years went by, they launched the UV-5R in 2012. The radio was packed with top-quality features and made two-way communication more versatile than ever before. Baofeng up till today continues to uphold its unwavering commitment of empowering radio enthusiasts with innovative and high-quali products that are packed with new features and launched at an affordable price.

Baofeng Radio Variants

Over the years, there have been a lot of different Baofeng radio variants that the company launched. These varian cater to your preferences regardless of whether you're a novice enthusiast or an expert. Each of the variants are packe with innovative features that improve the radio's overall functionality and enhance communication effectiveness. Th: said, let's dive into a few Baofeng options that may be suitable for you.

One of the first things that needs to be considered to understand the Baofeng radio is the naming. Before we get int this, it's possible that it may appear to be a bit difficult. However, following a step-by-step approach can simplify it. T understand the namings , let's start with the UV-5R.

The UV-5R is a radio that features a high-quality compact design. The radio is packed with innovative features an comes at an affordable price. A few common examples of some great features of the Baofeng UV-5R include its dua band capability, build quality and a simple and easy intuitive interface. These features make the UV-5R a great optio for beginners and experts and suitable for use in different scenarios such as emergency situations and gueril communication.

Another great variant that Baofeng has launched is the BF-F8HP. This radio comes with 8 watts of transmitting powe This capability alone makes it the perfect option for effective communication during adventure, emergency situation: and other times of crisis.

The UV-82HP is designed for those who seek maximum performance. It has an expanded frequency range, longer batter life, and is very portable. You can literally take this with you anywhere you go. These products captured the attentio of radio enthusiasts worldwide and influenced them in a way that's truly unique.

Lastly, the UV-82HP. This Baofeng radio is one that's designed for those who seek maximum performance. The UV 82HP has an expanded frequency range and longer battery life. These features and many others like them have mad this radio a popular option among newbie and expert enthusiasts alike.

Baofeng's Influence on Ham Radio Enthusiasts

The impact the Baofeng radio has had on ham radio enthusiasts can only be defined as profound. These radios remove the barriers to entry and made owning and operating such devices more accessible and simpler than before. The Baofen, radios feature an open-source firmware and are compatible with numerous third-party accessories and can be used b anyone regardless of their skill level.

Enthusiasts are known for owning unique devices and truly making them their own and that's something which can b easily done by the Baofeng radio. The customization options on the devices are truly endless. You can reprogram th software, browse and store frequencies as per your requirements and do so much more like creating a DIY antenna fo it, too.

he launch of the Baofeng radio and its popularity overtime have also led to the development of multiple online forums, cial media groups, and dedicated websites where like-minded individuals can engage with each other. These digital atforms serve as a great medium for accessing tips, troubleshooting advice, and projects for your Baofeng radio.

requency Uses and Applications

ow that we've covered the origins of the Baofeng radio, I'd like to share with you some of the different frequencies it pports and their applications.

ery High Frequency (VHF) Range: 136-174 MHz

- **Multi-Use Radio Service (MURS)**

  The MURS operates within the 151.820 to 154.600 MHz segment. By using this, you can have license-free access for personal or business communication which makes it an attractive option for short-distance, non-commercial uses.

- **Business Band VHF**

  The business band VHF frequencies such as 151.505 MHz and 154.540 MHz are part of the VHF business band, used for commercial operations. These channels require a license and are commonly used for internal business communications.

- **NOAA Weather Radio (162.400-162.550 MHz)**

  This VHF range includes frequencies that are specifically allocated for the NOAA Weather Radio service. By using these frequencies, you can have access to continuous weather and emergency alert information.

- **Marine VHF Radio**

  These frequencies are ones that are dedicated to marine communications and are vital for maritime navigation and safety.

- **2 Meter Amateur Radio Band (144-148 MHz)**

  This segment is popular among amateur radio enthusiasts for its effective balance of range and penetration and it can facilitate various communication forms from voice to digital modes.

Jltra-High Frequency (UHF) Range: 400-470 MHz

- **Family Radio Service (FRS) and General Mobile Radio Service (GMRS)**

  These services operate within the 462.5625 to 467.7125 MHz frequencies, supporting personal and family communications with varying power requirements and licensing rules.

- **UHF Business Band**

  Frequencies such as 464.500 MHz and 469.550 MHz are allocated for commercial use, essential for operational communications in environments where VHF is less effective.

- **70cm Amateur Radio Band (420-450 MHz)**

  This band is favored for its ability to support a wide range of communication methods, including satellite operations, due to its higher frequency and shorter wavelength.

Additional Frequency Range: 220-250 MHz (1.25m Band)

Some Baofeng models extend their coverage into this band. What this means for you is that you have access to a intermediary option between VHF and UHF. This band is less crowded and provides unique opportunities for communication experimentation and niche applications within the amateur radio community.

FM Broadcast Reception: 67-108 MHz

The Baofeng radio's ability to receive FM broadcasts is a valuable feature. You can use it to access music, news, and emergency information which allows you to use the device utility beyond simple two-way communications.

Security and Regulatory Considerations

Before using any of these frequencies, you need to know that each of them comes with a set of regulatory considerations to which you must adhere. These considerations are designed to optimize the spectrum usage and keep interference to a minimum. These regulatory requirements include necessary licenses and respecting designated channels for emergency service.

CHAPTER 2:
Setup and Initialization

y now, you're well aware of the origins of the Baofeng radio, the
oduct variations, frequencies and more. So, in this chapter, I'll share
ith you how you can set up and initialize the Baofeng radio. In this
apter, we'll start with the unboxing.

uring the unboxing you also need to assess all the components and
ake sure they are not broken or damaged. Towards the latter half of
is chapter, we'll also discuss all the features and buttons on your
aofeng radio. Let's begin.

Inboxing and Initial Assessment

eing a radio enthusiast myself, I know that there's nothing quite as
xciting as unboxing the radio. However, when you're going about
nboxing, it's crucial not to overlook anything and that all parts are
noroughly assessed. To do this, you can follow the step-by-step
pproach, I've outlined for you below:

1. **Inspecting the Packaging**

   When it comes to unboxing, the very first thing that you need
   to do is check the packaging. Sometimes during shipping, it's
   possible for transit related damages to occur. By checking the
   packaging, you can ensure that you don't have these issues with
   your radio as they can often lead to damaged components like
   antennas, belt clips, or other necessary accessories.

2. **Checking the Contents**

   The next thing you need to do is take inventory of all the components. The items that are supposed to be
   included are listed on the packaging so you need to make sure that none of them are missing. Some of these
   items include:

- The radio unit.
- A battery pack.
- An antenna.
- A belt clip.
- A charger.
- A user manual.

In addition to items, the package may also include other things that are specified by the manufacturer.

3. **Inspecting the Radio**

   When you're sure that you have all the components listed by the manufacturer or on the packaging the next thing you need to do is check the radio itself. You can start doing this by examining the exterior of the radio. What you need to look for here is signs of any physical damage.

   Common examples of such damages may include scratches, dents, or scuffs. When you are examining the radio the different parts that you need to check include the screen, buttons, and knobs. You must make sure that everything is intact and it's working properly

4. **Checking the Battery and Antenna**

   When you've made sure that the radio itself is in top notch condition the next thing you need to do is check the battery and the antenna. When examining the battery look at the contact points and make sure that they're free from any corrosion and absolutely clean.

   It's important for you to know that any damages to the contact points of the battery, regardless of its extent can impact the performance of your radio. You also need to check the antenna for any signs of damages and when you're putting it on the radio unit, make sure that you don't over tighten it as this damage the connector

5. **Powering On the Radio**

   Now that you have made sure that the radio and all its components are in working condition it's time to turn it on. It's important for you to know that with some models you may need to insert the battery back before you power up the radio. Another thing that you need to look for when turning on the radio is any startup message or indicators that are displayed on the screen.

By following the steps for the unboxing, you can make sure that you have all the components for the radio, none of them are damaged, and that the radio along with its components is functioning as required.

Attaching Battery and Antenna Components

The next thing that you need to do is assemble the radio and to do this you will have to attach the battery and the antenna. Now I understand that this may seem like a pretty simple thing to do, but there is actually a step-by-step process you need to follow. By doing so you can ensure that all the components are attached property and you can optimize your radio's performance. So, with that in mind, here what you need to do:

1. **Preparing the Battery Pack**

   Now, the very first thing that you need to do is prepare the battery pack for your radio. It's crucial for you to make sure that the battery is fully charged before you insert it. You can do this by using the charger provided by the manufacturer and remember that the battery should be charged up until the indicator shows that it's complete. After removing the battery from the charger make sure you check and inspect it for any science of damages or defects to the contact points.

2. **Inserting the Battery Pack**

   Once the battery is fully charged and it's not damaged you can insert it into the radio. The battery compartment on the radio is usually located on the backside. When you place the battery in the compartment make sure that the contact points on the battery align with ones on the radio. Once you turn that, press the battery until you hear a click sound.

3. **Attaching the Antenna**

   Next up is the antenna. This is one of the most critical components of the radio as it's responsible for sending and receiving audio signals. When you're connecting the antenna, the first thing that you need to do is look for the connector; it's usually located on the top of the radio unit. Once you've found the connector you need to align it with the connector and gently screw it in place.

4. **Securing the Antenna**

   After you attach the antenna, you need to make sure that it's securely fastened to the radio unit. To do this you can give the antenna a twist and make sure that it's snugly in place. However, it's important to remember that when you are doing this you cannot use too much force as it may damage the component.

5. **Verifying Connections**

   Now, after attaching the antenna and the battery you need to verify that both of these things are connected properly one final time. It's important to do this because it can help prevent any issues down the road and will help you prolong the lifespan of your radio. That's said, what you need to check here is any loose connections or irregularities and you need to make sure that everything is properly aligned.

Managing Antenna and Microphone Jacks

Now, the next thing we're going to talk about are the antenna and the microphone jacks. It's important for you to know that the Baofeng radio is equipped with a SMA-Male antenna connector. This connector is functional, but has its limitations. For instance, the connector is not durable under tough conditions. However, this should not limit the effectiveness of your communication efforts because you can switch to a BNC adapter instead. Using a BNC adapter has a ton of different benefits like:

- **Durability:** When it comes to durability, you need to know that a BNC adapter is designed to withstand a more stressful environment, so the risk of damaging it is lower.

- **Ease of Use:** The BNC adapter is also easier to use as it features a bayonet-style docking mechanism. What this means is that these adapters can be quickly attached or detached which makes changing the antenna a hassle free experience.

- **Compatibility:** The BNC adapter also increases the range of compatible antenna options for your Baofeng radi
With it you can easily use military grade and commercial antenna options that help improve the performance
your radio.

Now that you're well aware of the BNC adapter, here's what you need to do to use it.

1. **Purchase a BNC Adapter:** Look for a BNC adapter designed specifically for Baofeng radios or one that
compatible with SMA-Male to BNC-Female connections.

2. **Attach the Adapter to Your Radio:** Carefully thread the SMA-Male end of the adapter into your Baofeng
antenna port. Ensure it's securely fastened to avoid damage.

3. **Connect Your Antenna:** With the adapter in place, you can now connect any BNC-equipped antenna to you
radio. The bayonet locking mechanism should click into place, securing the antenna.

Features of the Baofeng UV-5R Radio

The next thing we're going to talk about in this book are the features and layout of the Baofeng radio. We'll look a
different things like the design elements and operational capabilities, all of which make it a popular option fo
communication needs in diverse circumstances. Covering these features, components, and capabilities is essential as i
will help you use the Baofeng radio more effectively. That said, let's look at each of them.

Antenna

The Baofeng radio comes with a removable antenna that can support sending and receiving radio signals across VHF (136-174 MHz) and UHF (400-470 MHz) frequencies. The antenna connection uses an SMA interface, which makes it easier to replace and upgrade if you want to improve the performance.

Flashlight

The radio also has a built-in LED flashlight that is located at the top of the radio. You can activate the flashlight with a quick press of the dedicated side key and it even has an additional mode for strobe lighting.

Volume Knob/Power Switch

The volume knob of the radio is located at the top and also serves as the power switch. You can turn the knob clockwise to power the radio and increase the volume. If you turn counterclockwise, it will decrease the volume and eventually power off the device.

LCD Display

The front panel of the Baofeng radio features an LCD display that shows operating information. Such information may include frequency/channel, battery level, signal strength, and activated functions (e.g., VOX, DCS/CTCSS codes).

Function Keys

As far as function keys are concerned, there are three things that you need to be aware of. These include:

- **Side Keys:** Two programmable side keys offer quick access to functions like the call/alarm feature and the flashlight/monitor function.
- **PTT Button:** The Push-To-Talk button, located on the side of the radio, initiates transmission when pressed.
- **Frequency/Channel Mode Switch (VFO/MR):** This button toggles between frequency mode (VFO) and memory channel mode (MR), allowing users to switch between direct frequency entry and preset channels.

LED Indicator

The LED indicator above the screen shows the radio's status. When it's red, it means that the radio is transmitting signals and when it's green it means that the radio is receiving signals.

Strap Buckle

The Baofeng radio also comes with a strap buckle that is used for attaching a wrist strap or lanyard, enhancing the radio's portability and security against drops.

Accessory Jack

The accessory jack on the radio is used as accommodation space for accessories such as headsets, speaker microphones, and programming cables. It uses a Kenwood two-prong connector standard. This jack serves two purposes that include providing support and programming the radio.

The jack supports a wide range of accessories, including external microphones, earpieces, and headsets. These accessories can enhance your radio's usability in various settings.

It can also be used for programming the radio. By connecting a programming cable from the jack to a computer, you can utilize software (such as CHIRP) to manage frequencies, settings, and channels more efficiently than manual programming allows.

Frequency Display Switches (A/B Key)

This button switches the focus between the top and bottom frequency/channel displays. It allows you to monitor and quickly switch between two frequencies or channels.

Band Switches (BAND Key)

The BAND key switches the radio between its supported frequency bands (VHF and UHF), facilitating easy access to different ranges of operation.

Keypad

The front-facing keypad enables direct entry of frequencies and access to the radio's menu for configuring various settings.

Speaker/Microphone

The front panel houses the speaker and microphone, allowing for clear audio reception and transmission.

Battery Pack

The UV-5R is powered by a rechargeable lithium-ion battery pack, which is easily removable for charging or replacement.

Battery Contacts and Removal Button

The battery pack attaches to the radio body via contacts, with a button provided to release the battery for removal or replacement.

Powering Up and Time Configuration

The next thing we're going to cover in this chapter is how you can power up the radio and configure the time correctly. To do that here's a list of steps to follow:

1. **Powering On the Radio**

   When you're powering on the Baofeng radio, you can find the power button on the top or side of the device. To power it on, you need to press and hold the button until the display lights up and you hear a beep. Once the radio is powered on you can release the button.

2. **Initial Setup Wizard**

When you turn on the radio for the first time you'll have to go through an initial setup wizard. This will allow you to set the language, region, and time zone. To complete the process, all you have to do is follow the on-screen prompts.

3. **Setting the Time**

Once the initial setup wizard is complete, you'll need to set the time on your Baofeng radio. To do this, the first thing that you need to do is access the time settings menu, usually found in the radio's settings or configuration menu. You can use the arrow keys or navigation buttons to adjust the time settings, including the hour, minute, and time zone.

4. **Using 24-Hour Time Format**

When it comes to the time format, you need to know that the Baofeng radio typically defaults to the 24-hour time format, also known as military time. If you prefer to use the 12-hour time format (AM/PM), you can usually find an option to switch between the two formats in the time settings menu.

5. **Syncing with Atomic Clocks**

Some Baofeng radios may have the option to sync the time with atomic clocks for increased accuracy. If this feature is available on your radio model, you can enable it in the time settings menu. The radio will automatically sync with atomic clocks via the built-in receiver to ensure precise timekeeping.

6. **Verifying Time Accuracy**

After setting the time on your Baofeng radio, you need to make sure that it's accurate. You should compare the displayed time on your radio with a reliable time source such as an atomic clock or digital watch. Now, make any necessary adjustments to ensure that the radio's time is synchronized and accurate.

ront Panel Programming

rogramming features that can be assessed from the front panel can help with a range of settings and functionalities. f you're someone who wants to use the Baofeng radio in changing circumstances, then learning how to do the front anel programming is essential. By doing so you can make on-the-go adjustments without using any computer software.

efore we dive into how you can use the front panel programming features, you need to know that in changing ircumstances your communication needs and protocols may become more dynamic. What this essentially means is hat you may need to use varying frequencies and channels to get your message across effectively. That said, here are ome of the most essential functions and programming steps that you need to be aware of.

Switching Between Frequency and Channel Modes

The VFO/MR button is used to toggle the radio between Frequency Mode (VFO) and Memory (Channel) Mode (MR). The Frequency Mode allows you to manually enter a specific frequency, while Channel Mode accesses pre-programmed channels.

Entering and Changing Frequencies

When you're in the Frequency Mode, you can simply use the keypad to enter the desired frequency. The Baofeng requires six digits for complete frequency entry, ensuring precision.

Locking the Keypad

To prevent accidental changes, you can lock the keypad by pressing and holding down the # key. When the keypad locked and lock icon appears on the display, and the keypad will be disabled except for the push-to-talk (PTT) button. You can unlock it by doing the same thing again.

Adjusting Squelch Levels

The squelch settings can help you filter out weak signals and background noise. To do this, you need to access the menu, then navigate to Menu 0 (Squelch), and from here you can set the level to 1 for minimal filtering. This allows the reception of the broadest range of signals.

Setting Frequency Steps

You can access Frequency steps through Menu 1. They determine the increment the frequency changes with each press of the arrow keys. A common setting is 6.25kHz, suitable for scanning and fine-tuning reception.

Transmit Power Adjustment

You can adjust the Transmit power levels via Menu 2 (TXP). This allows you to choose between Low, Medium (on certain models), and High settings to balance range and battery life according to operational needs.

Activating Voice-Operated Exchange (VOX)

For hands-free operation, you can enable VOX through Menu 4. By doing so, you can turn the radio into a voice activated device. This is especially useful for digital operations or when physical access to the radio is restricted.

Utilizing CTCSS/DCS Tones

To access specific repeaters or to establish a semi-private communication channel, you can reprogram the Continuous Tone-Coded Squelch System (CTCSS) and Digital-Coded Squelch (DCS) tones through Menu 10 - 13.

Repeater Offsets and Direction

For repeater use, you need to set the offset direction (Menu 25, SFT-D) and frequency offset value (Menu 26, Offset) according to the repeater's specifications. This way you can use the radio to transmit and receive on different frequencies, a common requirement for repeater access.

djusting Backlight and LED Settings

ou can also customize the Backlight and LED behavior through Menus 29-31, allowing operators to minimize visibility uring nighttime operations or to conserve battery life.

esetting the Radio

cases where you may need to reset the radio to its factory settings (either for troubleshooting or operational security), enu 40 provides the option you need.

Menu Customization

/ith Baofeng radio, you have access to a broad range of customization options you can use to optimize the performance f the device and to adjust it as per your preferences. These customization options will make the radio more useful for fferent tasks and environments.

o select one option, press MENU, select the desired function by entering the number or using the arrows, then press 1ENU again. After selecting the desired value, always press MENU to save.

he following table shows the most important customization options you have at your disposal.

Menu	Option	Purpose	Adjustment
0	Squelch Level	Filters background noise when no transmission is present	Select desired level (1-9) with UP/DOWN keys
2	Transmit Power	Balances communication range and operational security	Toggle between LOW, MID, or HIGH
3	Battery Save	Extends battery life by reducing power consumption when idle	Select OFF or save level (1-4)
4	VOX (Voice Operated Exchange)	Enables hands-free operation by activating transmission upon detecting speech	Select OFF or sensitivity level (1-10)
5	Wide/Narrow Band Selection	Adjusts bandwidth to match network standards or to optimize signal quality	Choose between WIDE or NARROW with UP/DOWN keys
7	TDR (Dual Watch)	Monitors two frequencies simultaneously	Select ON/OFF
8	Beep On/Off	Controls the beep sound for keypad presses	Select ON/OFF
9	Time-Out Timer (TOT)	Prevents accidental long transmissions by setting a maximum transmit duration	Choose duration with UP/DOWN keys

10-13	CTCSS/DCS Settings	Configures sub-audible tones for accessing repeaters and reducing interference	Select desired tone/code
14	Voice Prompt	Enables voice prompts for menu navigation and settings	Select ENG, CHI, or OFF
15	ANI ID	Sends caller ID for identification in group communications	Input ID
16	DTMFST (Dual-Tone Multi-Frequency Settings)	Configures the tone duration for DTMF dialing	Choose duration with UP/DOWN keys
17	S-CODE	Sets code for signaling and identification	Select code with UP/DOWN keys
19	PTT-ID	Sends an ID code each time the PTT button is pressed, for caller identification	Choose ID type with UP/DOWN keys
23	BCL (Busy Channel Lockout)	Prevents transmission on occupied frequencies	Toggle ON/OFF
39	End-of-Transmission Tone (ROGER)	Enables/Disables the beep heard at the end of a transmission	Toggle ON/OFF
40	Reset	Resets radio settings to factory default	Select ALL or VFO

CHAPTER 3:
Programming Your Baofeng Radio

Alright dear reader, by now you and I have covered how the Baofeng radio originated and what its essential components and functionalities are. But that's not all it's going to take to thrive as communication needs change. You need programming. Now this might seem a bit difficult at first but it's not. Let's begin!

Manual vs. Software-Based Programming

When it comes to programming your Baofeng radio, you have two primary options: manual programming and software-based programming. Now, the thing here is that both of the options are highly important. Each of them has its own benefits and drawbacks and you need to be familiar with both if you want to operate your Baofeng radio effectively.

Manual Programming

Manual programming involves taking care of the frequencies, offsets, tones, and other settings of the radio. For this, you'll need to use the keypad and the menu system. It's not as difficult as it seems and there are a lot of benefits, too. Some of them include:

1. **Independence**: Manual programming allows you to program your radio on the go. What this essentially means is that you won't need to rely on other equipment or software. This makes it ideal for situations where you need to quickly adjust your radio's settings without access to a computer. Think about it: you're stuck on a hiking trail and need to reach out to someone. You won't have access to software to reprogram your radio and that's when a manual approach comes in handy.

2. **Hands-On Learning**: Manual programming provides an excellent opportunity to familiarize yourself with your radio's menu structure and navigation buttons. You can gain a deeper understanding of your radio's operation by using this practical approach.

However, manual programming also has its limitations:

- **Time-Consuming**: Programming frequencies and settings manually can be time-consuming, especially if you nee to program multiple channels or complex configurations.

- **Potential for Error**: Manual programming increases the risk of input errors, such as mistyped frequencies incorrect settings. This can lead to communication issues or radio performance problems. Don't want any of tha to happen if you're stuck on a hiking trail, right?

Software-Based Programming

Software-based programming involves using dedicated programming software, such as CHIRP, to program your Baofen radio via a computer. I know this might be a bit technical, but it's not like programming where you have to code. Th method offers several advantages:

1. **Efficiency**: Software-based programming allows you to program multiple channels, frequencies, and settings usin a computer interface quickly and easily. This can save you time and effort compared to manual programming.

2. **Accuracy**: With software-based programming, you can view and edit radio settings in a user-friendly graphica interface, reducing the risk of input errors and ensuring accurate programming. It's actually pretty cool.

3. **Advanced Features**: Programming software often offers advanced features not available through manua programming, such as bulk editing, channel cloning, and frequency database integration. So, you basically have lot of different options at your disposal.

However, software-based programming also has some drawbacks:

- **Dependency on Equipment**: Software-based programming requires access to a computer, programming cable, an compatible software. If you're in the field without these resources, you may not be able to make programmin changes.

- **Learning Curve**: Using programming software may require a learning curve, especially for beginners. Familiarizin yourself with the software interface and understanding its features may take time and practice.

So, to wind it all up, both of the methods have their pros and cons. Which should you use? Well, to be honest, it depend on the situation you're in. If you're planning to head for an adventure and have all your communication needs figured ou beforehand, it's better to stick with the software programming methods. This way you'll be able to have everything se perfectly beforehand.

But this doesn't mean that you overlook the manual approach. Think about it: what if some unexpected needs arise? That' when you can use the manual programming methods.

Inputting Frequencies and Repeater Data

Alright, so that you're well versed with both the manual and software approaches and their pros and cons, let's look a what you need to do to input frequencies and repeater data. This is the very first step to customizing your device to mee your communication needs.

It allows you to access specific channels and utilize repeater systems for an extended communication range. By programming these settings, you can easily tune into the right frequencies and leverage repeaters to amplify your signal

learning how to master this, you can enhance your communication needs with the press of a few buttons and a couple clicks.

Entering or Changing a Frequency on the Baofeng Radio

you want to use your Baofeng radio like a pro, you need to know how to enter or change frequencies manually. Why? Well, if you're in an environment where the operational or communication requirements are changing with the blink of an eye, having the ability to manually program your radio gives some extra flexibility. I had a hard time with this at first, but I've listed down a few that'll help make things easier.

Switching Between Frequency and Channel Mode

1. **VFO/MR Button:** The first step in entering or changing frequencies is to determine whether the radio is set to Frequency Mode (VFO - Variable Frequency Oscillator) or Channel Mode (MR - Memory Recall). Press the "VFO/MR" button located below the radio's display screen. If the voice prompt feature is enabled, the radio will audibly indicate the current mode as either "Channel Mode" or "Frequency Mode". In Frequency Mode, you can directly input a new frequency; in Channel Mode, you can select from pre-programmed channels.

Entering a New Frequency

2. **Entering Frequency Mode:** Ensure the radio is in Frequency Mode (VFO). If not, press the "VFO/MR" button to switch to it. The display will show frequency numbers without any channel indicators.

3. **Inputting the Frequency:** Directly enter the desired frequency using the keypad. For instance, to set the radio to 146.000 MHz, press the keys "1", "4", "6", "0", "0", "0" in sequence. It's important to input all six digits of the frequency, including trailing zeros, to complete the entry.

4. **Confirmation:** After entering the frequency, it should now be displayed on the LCD screen. There's no need to press an "Enter" key; the radio automatically updates to the newly input frequency.

Correcting Mistakes

5. **If an Error Occurs:** Should you make a mistake while entering the frequency, you can either press the "EXIT" button to cancel the input or simply pause for a moment. The radio will automatically revert to the previous frequency if no further input is detected within a few seconds.

Switching Frequencies Between Display Areas

6. **A/B Button:** The Baofeng radio features dual watch functionality, allowing two frequencies to be monitored simultaneously, one on the top display and one on the bottom. Press the "A/B" button to toggle which frequency (top or bottom) you're actively controlling or changing. An arrow icon on the display indicates the currently active frequency.

Program Repeater Offset: If you're programming a repeater frequency, you'll need to set the offset. To do this, you need to navigate to the repeater offset menu and select the appropriate offset direction (+ or -) and frequency shift (typically 00 kHz for VHF and 5.000 MHz for UHF).

Enter CTCSS/DCS Tone: If required, input the CTCSS (Continuous Tone-Coded Squelch System) or DCS (Digital-Code Squelch) tone for access to repeater systems or to filter unwanted transmissions.

Save Settings: Once you've entered all the right frequencies and settings, save the configuration by pressing the **Me** button and selecting the save option.

Test Communication: After programming, test your radio by tuning into the programmed frequencies and attempting communicate. During the testing phase, make sure that you can access repeater systems and communicate effectively.

Saving and Organizing Channels

Saving and organizing channels on your Baofeng radio is a pivotal step in customizing your device to suit you communication needs. Saving and organizing channels enables you to quickly access your preferred frequencies and settings without the hassle of manual input each time.

By organizing channels, you can categorize them based on frequency bands, usage scenarios, or geographical location allowing for streamlined operation and enhanced communication efficiency. Now, I know that you might be thinking tha saving a channel is a pretty straightforward process so what's the need for a tutorial?

Well, the process is pretty straight forward, but there are a couple of steps involved. So, with that in mind, here's wha you need to do:

1. **Access Channel Menu**: Press the **Menu** button on your Baofeng radio to access the channel programming menu.

2. **Select Memory Channel**: Using the arrow keys, navigate to the desired memory channel slot where you want t save the channel settings.

3. **Enter Frequency**: Input the frequency of the channel using the keypad. For example, to enter a frequency of 146.52 MHz, press **1-4-6-5-2-0**.

4. **Save Channel**: Once the frequency is entered, press the **Menu** button again to save the channel to the selecte memory slot.

5. **Edit Channel Name (Optional)**: If desired, you can edit the channel name to provide a clear label for eas identification. Navigate to the channel name editing menu, use the keypad to enter the desired name, and pres **Menu** to save.

6. **Organize Channels**: After saving channels, you can organize them by grouping similar frequencies together. Creat folders or categories based on frequency bands, usage scenarios, or geographical locations to facilitate eas navigation.

7. **Assign Channel Priority (Optional)**: For frequently used channels, consider assigning priority status to ensur quick access during emergencies or critical communications.

8. **Test Channels**: Once channels are saved and organized, test each channel to ensure proper functionality. Verif that you can tune into the programmed frequencies and communicate effectively.

CHIRP for Baofeng Programming

CHIRP is a powerful programming tool that simplifies the process. With its user-friendly interface and robust features, CHIRP is a must-have tool for Baofeng enthusiasts looking to customize their radios with ease. It has an intuitive user interface and robust features and believe me, it's an essential tool if you want to customize your radio.

CHIRP allows users to program Baofeng radios using a computer, offering several advantages over manual programming:

- **Efficiency**: CHIRP streamlines the programming process, allowing users to input frequencies, channels, and settings quickly and accurately. This saves time and reduces the risk of errors associated with manual programming.

- **Ease of Use**: The intuitive interface of CHIRP makes it accessible to users of all experience levels. Its menu-driven design simplifies navigation, making it easy to find and adjust radio settings.

- **Advanced Features**: CHIRP offers advanced features not available through manual programming, such as bulk editing, cloning, and importing/exporting channel lists. These features enable users to customize their radios more comprehensively.

Getting Started with CHIRP

Now that you're familiar with CHIRP and what you can do with it, here's what you need to do to get started.

1. **Download and Install CHIRP**: Visit the CHIRP website and download the latest version of the software compatible with your operating system. Follow the on-screen instructions to install CHIRP on your computer.

2. **Connect Your Baofeng Radio**: Use a compatible programming cable to connect your Baofeng radio to your computer. Ensure that the cable is securely connected to both the radio and the USB port on your computer.

3. **Launch CHIRP**: Open the CHIRP software on your computer. Once opened, select the option to create a new radio configuration or open an existing configuration file.

4. **Read from Radio**: Click on the "Radio" menu and select the "Download From Radio" option. Follow the prompts to read the current configuration from your Baofeng radio into CHIRP.

5. **Program Frequencies and Settings**: Use the intuitive interface of CHIRP to input frequencies, channels, and settings according to your preferences. You can also import frequencies from external sources or clone settings from another radio.

6. **Write to Radio**: After programming your desired channels and settings, click on the "Radio" menu and select the "Upload To Radio" option. Follow the prompts to write the configuration from CHIRP to your Baofeng radio.

And that's all! See how simple and easy it is to program new frequencies and channels to your Baofeng radio. All you have to do is follow the steps above and you're good to go.

CHAPTER 4:
Baofeng's Operating Modes

We're now going to cover operating modes. Learning about these modes can be a bit complex, but it's necessary. So, I'll save you some time and give you just of it right away. The Baofeng radio has a lot of different operating modes that you can choose from and each one serves a different purpose.

With each different mode, you either access unrestricted radio frequencies, store your preferred channels, or monitor and communicate on multiple frequencies. All of this can be quite handy when it comes to tailoring your radio to varying communication needs. So, with that in mind, let's look at the first operating mode.

Variable Frequency Operation (VFO) Mode

The VFO operating mode is a truly versatile feature that allows you to tune in to any frequency within your radio supported range. It allows you to have unrestricted access to the entire frequency spectrum. Unlike Memory mode, where channels are pre-programmed and fixed, VFO mode enables you to tune to specific frequencies on the go. This makes it ideal for exploring new frequencies through simplex communications, or accessing repeater inputs. Now, let's look at how you can use this operating mode.

Navigating VFO Mode

With VFO mode, the airwaves become your playground, offering endless opportunities for exploration and communication. Whether you're scanning for distant stations, searching for local repeaters, or participating in

mateur radio contests, VFO mode empowers you to take full control of your Baofeng radio and unlock its true potential.

y mastering Variable Frequency Operation in VFO mode, you gain the freedom to tune into any frequency at any time, pening up a world of possibilities for amateur radio enthusiasts and communication enthusiasts alike.

he VFO mode might seem a bit complicated at first. But, believe me, using it is very simple. Just follow the steps below nd you'll be good to go.

1. **Access VFO Mode**: Press the **VFO/MR** button on your Baofeng radio to switch to Variable Frequency Operation mode.

2. **Select Frequency Band**: Use the arrow keys or the tuning knob to select the desired frequency band, such as VHF or UHF.

3. **Enter Frequency**: Input the desired frequency using the keypad. For example, to tune to 146.520 MHz, enter **1-4-6-5-2-0**.

4. **Fine-Tune Frequency**: Use the tuning knob to fine-tune the frequency to your exact preference. You can adjust the frequency by rotating the knob in a clockwise or counterclockwise direction.

5. **Monitor Activity**: Once tuned to the desired frequency, monitor for activity by listening for transmissions or using the radio's signal meter to gauge signal strength.

6. **Engage in Communication**: When you find a channel of interest, engage in communication by keying the microphone and transmitting your message. Remember to follow proper radio etiquette and regulations.

Channel Access through Memory Mode

Memory mode is a feature that's designed to help you store and access frequencies and channels with ease and makes sing the Baofeng radio very convenient. This operating mode comes in handy because you don't really need to enter he frequencies manually. You can just access memory mode and choose one within seconds.

Navigating Memory Mode And Programing Channels

Whether you're coordinating with fellow operators, monitoring emergency frequencies, or accessing repeater networks, Memory Mode ensures that your essential channels are right at your fingertips. So, with that in mind let's look at how ou can navigate and program channels to Memory Mode. For navigating:

1. **Access Memory Mode**: Press the **VFO/MR** button on your Baofeng radio to switch to Memory Mode.

2. **Select Memory Channel**: Use the arrow keys or the tuning knob to scroll through the stored memory channels.

3. **Retrieve Stored Frequency**: Once you've located the desired memory channel, press the **A/B** button to access the stored frequency.

4. **Monitor Activity**: Listen for activity on the selected memory channel by adjusting the volume and squelching settings as needed.

5. **Engage in Communication**: When ready to transmit, key the microphone and deliver your message. Remember to adhere to proper radio etiquette and regulations.

The process for programming channels in memory mode is a bit different. Here's what you need to do:

1. **Enter Programming Mode**: Press the **VFO/MR** button to access Memory Mode, then press the **MENU** butto followed by the **27** button to enter programming mode.

2. **Select Memory Channel**: Choose the memory channel slot where you want to store the frequency.

3. **Enter Frequency**: Input the desired frequency using the keypad.

4. **Save Frequency**: Press the **MENU** button followed by the **27** button again to save the programmed frequenc to the selected memory channel.

5. **Label Memory Channel**: Optionally, assign a name or label to the memory channel for easy identification.

Scanning Techniques

Alright, now let's move on to scanning techniques. To start off, I'll let you know that these techniques are a great wa to monitor multiple frequencies and channels. Believe me this feature comes in quite handy when you're stranded in remote area either by accident or because you choose to go there for an adventure.

When you use this feature, you can search for nearby conversations, check emergency channels or scan through repeate networks. All of this helps you become more aware of the situation around you and helps you communicate mor effectively.

Basic Scanning Operations And Advanced Technique

When you're trying out this feature, I urge you to try different techniques and parameters. This can help you fin frequencies and channels better suited to your needs. So, with that in mind, first let's look at how you can perform som basic scanning operations.

1. **Activate Scan Mode**: Press the **SCAN** button on your Baofeng radio to initiate scanning mode.

2. **Select Scan Parameters**: Use the keypad to enter the desired scan parameters, such as frequency range o channel selection.

3. **Start Scanning**: Press the **SCAN** button again to begin scanning through the specified frequencies or channels

4. **Monitor Activity**: As the radio scans, listen for activity on each frequency or channel. The radio will sto scanning when activity is detected.

5. **Respond to Activity**: If you hear activity on a particular frequency or channel, you can choose to engage i communication by stopping the scan and tuning to that frequency.

Once you've gotten set with the basics, you can move over to the advanced scanning techniques. Here's what thes techniques are and how you can use them.

- **Priority Scan**: Prioritize specific channels or frequencies for scanning by assigning them to priority channels Press the **PRI** button to activate priority scanning.

- **Dual Watch**: Monitor two channels simultaneously with dual watch mode. Set the main channel and sub channel frequencies, then activate dual watch to monitor both frequencies simultaneously.

- **Skip Channels**: Exclude certain channels or frequencies from the scanning sequence by programming them a skipped channels. This allows you to focus on relevant frequencies without interruptions.

- **Delay Function**: Customize the scan delay to control how long the radio pauses on each channel before moving to the next. Adjusting the delay time ensures you have sufficient time to listen for activity on each channel.

Dual Listening with Dual PTT Features

Thought that the Baofeng radio couldn't get any better? Well think again! Its dual Push-to-Talk (PTT) feature is truly amazing. By using this feature, you can monitor and transmit to multiple channels at the same time, just as the name suggests. Believe it or not, but this neat little feature can be the difference maker.

It enhances your awareness and versatility in communication scenarios where you need to be engaged on multiple channels. This can happen when you're working with multiple teams, monitoring emergency frequencies, and when you're just trying to keep regular communication running smoothly. Some of its distinctive advantages include:

- Enhanced Situational Awareness: Monitor multiple channels simultaneously to stay informed about developments across different communication networks.
- Improved Coordination: Coordinate effectively with multiple teams or groups by monitoring their communication channels while maintaining communication on your primary channel.
- Quick Response: Respond promptly to incoming messages or calls on either channel without the need to switch back and forth between channels.
- Versatile Communication: Adapt to changing communication needs by seamlessly switching between channels as required, ensuring effective communication in dynamic environments.

Activating Dual Listening Mode

Now that you're aware of what PTT is all about and how it can benefit you, just follow the three simple steps below to use it:

1. **Activate Dual PTT**: Press and hold the **DUAL** button on your Baofeng radio to activate dual listening mode.
2. **Select Main and Sub Channels**: Choose the main and sub channels you wish to monitor and transmit on simultaneously. Use the keypad to enter the frequencies or channel numbers.
3. **Engage in Communication**: With dual listening mode activated, you can now monitor activity on both channels and transmit on either channel using the corresponding PTT buttons.

CHAPTER 5:
CTCSS and DCS Signal Tones

Now that you're familiar with the different operating modes on your Baofeng radio and how you can use each of them, let's shift our focus to learning about signal tones. Based on what I have found, most people tend to overlook this aspect when they're learning about Baofeng radio.

But, believe me, it's one that you need to cover in detail. By learning how to use different signal tones, you can ensure selective filtering, reduce interference, have private conversations, and improve signal clarity. Pretty fascinating, isn't it? Let's begin!

CTCSS Tones

Continuous Tone-Coded Squelch System (CTCSS) tones play a very important role. They can help you enhance privacy and reduce interference, meaning that you can streamline your radio communications with ease. Now you're probably wondering how these tones allow you to do that.

Well, allow me to explain. You see these are sub-audible tones that are transmitted alongside what you're saying. What this means is that if a receiver is not set to the same CTCSS tone, your transmission will not be heard, meaning that you're essentially implementing a selective filter over your radio transmission.

But that's not all CTCSS tones are good for. They have a few other benefits that include:

- Enhanced Privacy: Ensure private communication by transmitting and receiving signals only from radios set to the same CTCSS tone, preventing unauthorized listeners from eavesdropping on your conversations.

- Reduced Interference: Minimize interference from other users or external sources operating on the same frequency, allowing for clearer communication even in congested radio environments.
- Selective Reception: Selectively filter incoming transmissions based on their CTCSS tones, allowing you to focus on relevant communications while disregarding irrelevant or unwanted signals.

- Improved Channel Sharing: Facilitate efficient channel sharing among multiple users by assigning unique CTCSS tones to different groups or individuals, enabling simultaneous communication on the same frequency without interference.

Programming CTCSS Tones

Now that you know what CTCSS tones are and how they can benefit you, it's time we learn about using them, too. The process is incredibly simple. Just follow the steps below and you're set.

1. **Access Menu**: Press the **MENU** button on your Baofeng radio to enter the menu.
2. **Select CTCSS Function**: Navigate to the CTCSS function by using the arrow keys.
3. **Choose Desired Tone**: Scroll through the list of available CTCSS tones using the arrow keys and select the desired tone for your channel.
4. **Confirm Selection**: Press the **MENU** button again to confirm your selection and exit the menu.

Optimizing CTCSS Tone Selection

With CTCSS tones at your disposal, you possess the tools to communicate effectively, privately, and seamlessly in any radio environment. The one thing I want you to remember is that when you optimize CTCSS tones, it's very important to experiment as it can help you find the right combination for your communication needs. When you're experimenting, consider different factors such as the level of background noise, the presence of other users on the same frequency, and the desired level of privacy to tailor your CTCSS settings accordingly/

DCS Mechanisms

Digital-Coded Squelch (DCS) mechanisms are a way to increase communication privacy, decrease interference, and optimize signal reception. Now you might think that this is the same thing as CTCSS tones, but it's not. You see, DCS tones use unique digital codes to optimize signal communication and improve signal reception. The four key benefits of DCS mechanisms include:

- Enhanced Communication Privacy: Encrypt radio transmissions with unique digital codes, ensuring that only receivers programmed with the same DCS code can decode and understand the transmitted signals, thereby safeguarding communication privacy.
- Interference Reduction: Minimize interference from other users or external sources by employing DCS squelch mechanisms, which filter out signals that do not match the programmed digital code, allowing for clearer and more reliable communication.

- Selective Reception: Selectively filter incoming transmissions based on their DCS codes, enabling users to focus on relevant communications while disregarding irrelevant or unwanted signals, thereby optimizing channel utilization and enhancing communication efficiency.

- Signal Reliability: Improve signal reception and minimize false squelch openings by utilizing DCS technology which provides a more robust and reliable means of decoding signals in noisy or congested radio environments.

DCS Configuration

Setting up the DCS configuration protocols is quite simple and all you have to do is follow the four steps I've mentioned below. That's it.

1. **Access Menu**: Press the **MENU** button on your Baofeng radio to enter the menu.
2. **Select DCS Function**: Navigate to the DCS function by using the arrow keys.
3. **Choose Desired Code**: Scroll through the list of available DCS codes using the arrow keys and select the desired code for your channel.
4. **Confirm Selection**: Press the **MENU** button again to confirm your selection and exit the menu.

Optimizing DCS Code Selection

As far as optimizing DCS codes is concerned, you can do that by following the same process that I shared with you for CTCSS tones. Just as a reminder, you have to experiment with different codes and determine which ones suit you best while considering factors like background noise, communication privacy, and the presence of the same frequency.

Configuration and Application of Tones

Configuring and applying tones is just as much art as it is science. That's exactly why we're going to see how we can configure and apply tones. But before I share the steps with you there's something that you need to be aware of.

You see the fundamental idea behind configuring and applying tones is that you set specific tones to different channels and frequencies so that you can meet your communication needs. To configure different tones, here's what you need to do:

1. **Access Menu**: Press the **MENU** button on your Baofeng radio to enter the menu.
2. **Select CTCSS Function**: Navigate to the CTCSS function by using the arrow keys.
3. **Choose Desired Tone**: Scroll through the list of available CTCSS tones using the arrow keys and select the desired tone for your channel.
4. **Confirm Selection**: Press the **MENU** button again to confirm your selection and exit the menu.

Now, once you've configured different CTCSS tones, you can also apply DCS tons by following the steps below:

1. **Access Menu**: Press the **MENU** button on your Baofeng radio to enter the menu.
2. **Select DCS Function**: Navigate to the DCS function by using the arrow keys.
3. **Choose Desired Code**: Scroll through the list of available DCS codes using the arrow keys and select the desired code for your channel.

4. **Confirm Selection**: Press the **MENU** button again to confirm your selection and exit the menu.

Ensuring Interference-Free Communication

One last thing I want to share with you before we wind things up is how you can ensure interference-free communication. You see, when you have inference in tones and signals, it limits your ability to use the Baofeng radio to communicate effectively.

Think about it, if you're trying to reach out to a receiver in an emergency situation, having the signal drop or clash with others every second isn't going to help, is it? That's why you need to mitigate interference. However, to that you first need to identify it and that can be done by following the steps below:

1. **Perform Frequency Scan**: Press the **MENU** button on your Baofeng radio and navigate to the frequency scanning function.

2. **Initiate Scan**: Enter the command **'MENU' > 'Scan' > 'Start'** to initiate the frequency scan process.

3. **Analyze Results**: Review the scan results displayed on the screen to identify active frequencies and potential sources of interference.

Now, when you've identified where the interference is, it's time to mitigate it. To do that you need to:

1. **Enable CTCSS/DCS Tones**: Assign CTCSS or DCS tones to your channels to filter out unwanted transmissions and minimize interference from other users or external sources.

2. **Adjust Squelch Settings**: Use the squelch function to suppress background noise and eliminate weak or unwanted signals, ensuring clear communication on your desired frequency.

3. **Select Clear Channels**: Based on the frequency scan results, manually select clear channels with minimal interference for your communication needs.

Optimizing Antenna Placement

Once you've done that you also need to optimize the antenna placement. So, you need to make sure that the antenna is securely attached to your Baofeng radio and that it's positioned vertically. Then, make sure that your radio is placed in an open area where there are no buildings, trees, or other objects that may cause interference. This last step might seem a bit obvious, but is one that is overlooked, but based on my experiences, placing your Baofeng radio in an open area works wonders when it comes to minimizing interference!

CHAPTER 6:
Baofeng's Advanced Functions

We've learned quite a lot about the Baofeng radio up till now, but there's still lots of ground to cover. In this chapter, we'll talk about all the advanced functions of the Baofeng radio. This includes voice activation , precision management, personalizing the display, and more.

For each function, I'll share with you its importance, functionality, how you can activate it, and the benefits you'll get from using it. So, without further ado, let's begin.

Voice Activation with VOX Technology

The concept of voice activation, as the name suggests, is quite simple. The Baofeng radio uses VOX technology to give you a hands-free communication experience. It automatically activates the microphone based on your voice command, meaning that you don't really need to push that PTT button when you have the VOX mode on.

The activation process is about as simple as it can get. But, once you're done with that, you need to adjust the sensitivity and test the voice command functionality as well. Starting to break a sweat? No need. Just follow the steps I've laid out for you, and you'll be all set for your next big adventure.

Activating VOX Mode

    a. **Access Menu Settings**: Press the **MENU** button on your Baofeng radio to enter the menu mode.

    b. **Navigate to VOX Settings**: Scroll through the menu options using the arrow keys until you find the VOX settings.

    c. **Enable VOX Mode**: Select the VOX mode and press the **MENU** button to enable VOX functionality.

djusting VOX Sensitivity

d. **Enter VOX Sensitivity Settings**: Access the VOX sensitivity settings within the menu options.

e. **Adjust Sensitivity Level**: Increase or decrease the sensitivity level according to your preference and environmental conditions. Higher sensitivity allows for activation with softer sounds, while lower sensitivity requires louder input.

f. **Save Changes**: Once you've adjusted the sensitivity level, save the changes to apply the new settings.

esting VOX Functionality

g. **Activate VOX Mode**: With VOX enabled, speak into the microphone to test the functionality. The radio should transmit your voice automatically without the need to press the PTT button.

h. **Adjust Sensitivity if Necessary**: If the VOX mode does not activate or is too sensitive, adjust the sensitivity settings accordingly and repeat the test until optimal performance is achieved.

enefits of VOX Technology

Having the voice command functionality activated will come in quite handy. Think about it: you don't need to reach or the radio. Just use your voice to activate the radio and you're all set to communicate. But that's not all. Here are ome other great benefits:

1. **Hands-Free Operation**: VOX technology enables hands-free communication, allowing users to multitask and maintain situational awareness while transmitting messages.

2. **Enhanced Convenience**: By eliminating the need to press the PTT button, VOX technology streamlines communication processes, particularly in dynamic environments where manual operation may be impractical.

3. **Improved Safety**: In situations where hands-free communication is essential for safety reasons, such as while driving or operating equipment, VOX technology ensures that users can stay connected without distractions.

Precision Measurement with TDR Functionality

Next up on our list of the Baofeng radio's advanced functions is the Time-Domain Reflectometer (TDR) functionality. By the name, it looks like it has something to do with a camera lens, doesn't time. But that's not the case. You see, the TDR functionality is what allows you to troubleshoot antenna issues with great precision.

It allows you to measure the impedance and identify faults along transmission lines. The TDR technology basically analyzes the time it takes for a signal to reflect back from the end of the cable. By using this functionality, you can gain valuable insights into the condition and performance of the transmission line.

So, now that you have a basic understanding of what the TDR functionality is, let's look at how you can activate it, perform measurements, and interpret graphs. This might sound a bit technical at first, but here's all that you need to do:

1. **Activating TDR Mode**

a. **Access Menu Settings**: Press the **MENU** button on your Baofeng radio to enter the menu mode.

b. **Navigate to TDR Settings**: Scroll through the menu options using the arrow keys until you find th TDR settings.

c. **Enable TDR Mode**: Select the TDR mode and press the **MENU** button to activate TDR functionality

2. **Performing TDR Measurements**

a. **Connect TDR Probe**: Attach the TDR probe to the end of the transmission line (e.g., coaxial cable) th you want to measure.

b. **Initiate Measurement**: Press the **TDR** button on your Baofeng radio to initiate the measuremer process.

c. **Analyze Results**: The radio will display a graphical representation of the impedance along th transmission line, allowing you to identify impedance mismatches and locate faults such as open or shor circuits.

Interpreting TDR Graphs

d. **Impedance Mismatch**: A sudden change in impedance indicated by a spike or dip in the graph may signif an impedance mismatch or discontinuity in the transmission line.

e. **Fault Detection**: An abrupt change in the slope of the graph may indicate the presence of a fault, suc as an open or short circuit, at that point along the transmission line.

Benefits of TDR Technology

Learning how to use the TDR functionality is very important as it allows you to ensure optimal performance and ca improve the overall effectiveness of your communication efforts. Some of the key benefits of the TDR include:

1. **Accurate Troubleshooting**: TDR technology provides precise measurements of impedance and identifies fault along transmission lines, enabling users to troubleshoot antenna and cable issues with confidence.

2. **Efficient Maintenance**: By pinpointing the location of faults, TDR functionality streamlines maintenance task and reduces downtime associated with diagnosing and repairing transmission line problems.

3. **Enhanced Performance**: With the ability to optimize antenna and cable configurations based on TDI measurements, users can maximize the performance and reliability of their radio systems.

Personalizing Display Settings

You already know that the Baofeng radio has a lot of different customization options for receiving and transmittin audio. From signals to tones, you're in complete control. But the interesting thing about the Baofeng radio is that it ha customizations options for visuals as well. You can personalize the display settings by adjusting a wide variety o parameters.

Common examples of such parameters include display contract, backlight duration, and colors and more. What thi means is that you can tailor the interface to your preferences and enhance the experience which makes using the Baofeng radio in different environments much easier. To do this, just follow the series of steps mentioned below:

1. **Accessing Display and Illumination Settings**

    a. **Enter Menu Mode**: Press the **MENU** button on your Baofeng radio to access the menu mode.

    b. **Navigate to Display Settings**: Scroll through the menu options using the arrow keys until you find the display settings submenu.

    c. **Adjust Display Parameters**: Select the desired display parameters, such as contrast and backlight duration, and use the arrow keys to adjust the settings to your preference.

    d. **Navigate to Illumination Settings**: Scroll further down the menu options until you find the illumination settings submenu.

    e. **Customize Illumination Colors**: Select the illumination color option and choose from the available color options to customize the backlight and keypad illumination to your liking.

2. **Enhancing User Experience**

    a. **Optimizing Visibility**: Adjusting display contrast ensures optimal visibility of screen content under different lighting conditions, such as bright sunlight or low-light environments.

    b. **Conserving Battery Life**: Setting backlight duration allows users to conserve battery life by adjusting how long the backlight remains active after a button press or interaction.

    c. **Personalizing Aesthetics**: Choosing illumination colors allows users to personalize the appearance of their Baofeng radio, adding a touch of style and flair to their device.

3. **Creating the Perfect User Interface**

    a. **Balance of Form and Function**: Customize display and illumination settings to strike the perfect balance between aesthetics and usability, creating a user interface that is both visually appealing and functional.

    b. **Tailored to Your Needs**: Adjust settings according to your specific requirements and preferences, whether you prefer a high-contrast display for outdoor use or soothing illumination colors for nighttime operation.

    c. **Reflecting Your Personality**: Personalize your Baofeng radio to reflect your personality and style, transforming it into a unique and distinctive device that stands out from the crowd.

Dual PTT – Push-to-Talk Operations

know that we've already covered one before, but here, we're going to look at it from a different angle. By allowing you to monitor and transmit on multiple channels, the dual PTT function helps you improve the effectiveness of your communications in diverse situations.

These situations could be a camping trip, an emergency, daily communication and a number of other things. Now, you already know how to activate the dual PTT function, so we won't waste time going over that again. That said, let's quickly go over how the dual PTT function can make your communication efforts more effective.

1. Efficient Team Coordination: Dual PTT enables seamless communication between multiple teams or groups operating on different channels, facilitating efficient coordination and collaboration in dynamic environments.

2.   Expanded Situational Awareness: By monitoring and transmitting on multiple channels simultaneously, use gain enhanced situational awareness and can respond rapidly to changing circumstances, ensuring effecti communication and decision-making.

3.   Versatile Application: From emergency response operations to outdoor adventures and recreational activitie Dual PTT offers unmatched versatility, allowing users to adapt their communication strategy to meet th demands of any situation.

4.   Unprecedented Control: With Dual PTT, users have unprecedented control over their communication device enabling them to transmit on multiple channels with ease and precision.

5.   Maximizing Efficiency: By streamlining push-to-talk operations and enabling simultaneous transmission c multiple channels, Dual PTT maximizes communication efficiency, ensuring clear and reliable communicatio in any scenario.

6.   Elevating Communication Experience: Experience the next evolution of push-to-talk communication with Du. PTT, a feature that redefines the way you communicate and empowers you to stay connected, informed, and i control at all times.

CHAPTER 7:
Emergency Readiness Strategies

o far, you've probably figured that the Baofeng radio is a reat tool to have if you're going on an adventure simply ecause it makes your communication with the outside orld much more effective. But the Baofeng radio is not ist good for adventures. It's equally beneficial, if not more, emergency situations, too.

y Using the Baofeng radio, you can access emergency hannels, get weather alerts, create emergency plans, carry ut search and rescue missions, and do so much more. The eatures might not look like much s at first, but they are ctually quite handy. Take the weather alert feature for xample.

et's say that you're on a camping trip and want to go for a ike on a nearby trail. Now, it's obvious that you're unaware f the weather or how it will change over the next few ours. If you move forward with such an approach, you'll likely be unprepared for any changes in the weather that do ccur.

Iowever, if you have the weather alert feature on your Baofeng radio turned on, it will predict weather patterns and rovide information about the changing conditions which you can then use to tailor your plans. And remember, this is just ne example.

hroughout this chapter we'll talk more about how the Baofeng radio can be of use in different emergency situations. Let's egin!

Emergency Channels and Frequencies

The Baofeng radio allows you to tune into different emergency channels and frequencies. By using these channels and frequencies you can have communication access to first responders, emergency services, and the general public. What important to know about the channels is that they're designed for rapid responses in times of crisis. Let's go back to our camping trip example for a second.

Now imagine that you or someone who needs medical assistance, but you're running low on first-aid supplies. That exactly when these channels will come in handy. They'll allow you to broadcast alerts and can play an important role in safeguarding lives. All of this essentially allows you to be more prepared and resilient during emergencies in a number of different ways that include:

1. Stay Informed: By accessing emergency channels and frequencies, you stay informed about potential threat severe weather conditions, and unfolding emergencies in your area, allowing you to take timely action to ensure your safety and well-being.

2. Coordinate Response Efforts: In times of crisis, emergency channels serve as vital communication links for coordinating response efforts, enabling first responders, emergency services, and community members to work together seamlessly to mitigate the impact of disasters and save lives.

3. Access Assistance: In the event of an emergency, accessing designated emergency channels and frequencies allow you to call for assistance, request help, and communicate critical information to responders and authorities ensuring prompt assistance and effective response.

4. Take Control: With knowledge of essential emergency channels and frequencies, you take control of your safety and well-being, ensuring you're equipped to respond effectively to emergencies and navigate challenging situation with confidence and resilience.

5. Prepare for the Unexpected: By familiarizing yourself with emergency communication protocols and accessing designated channels and frequencies, you prepare yourself for the unexpected, empowering yourself to respond calmly and effectively to emergencies as they arise.

6. Stay Connected, Stay Safe: In an uncertain world, access to essential emergency channels and frequencies keep you connected and informed, providing a lifeline to safety and security when you need it most.

These channels or frequencies are critical and you can use them, when needed, by following the three simple step mentioned below:

1. **Activate Emergency Mode**: Enter the menu of your Baofeng radio and navigate to the emergency settings. Enable emergency mode to access dedicated emergency channels and frequencies.

2. **Tune into Emergency Channels**: Using the keypad or tuning knob, navigate to the designated emergency channel on your Baofeng radio. Consult local emergency communication guides to identify the appropriate frequencies for your region.

3. **Monitor for Critical Alerts**: Once tuned into emergency channels, monitor for critical alerts, weather updates and emergency broadcasts. Stay vigilant and informed to respond effectively to evolving situations.

Weather Alert Capabilities

Next up on our list is weather alert capabilities. Along with emergency channels and frequencies, the Baofeng radio comes equipped with capabilities that are designed to provide notifications about severe or changing weather conditions in a timely manner. You can use these features and stay proactive in a number of different ways that include:

1. Stay Vigilant: Keep your Baofeng radio powered on and tuned to NOAA weather channels to monitor incoming weather alerts. Remain vigilant, especially during periods of inclement weather, and be prepared to take action based on the information provided.

2. Act Promptly: Upon receiving a weather alert, assess the situation and take appropriate precautions to ensure your safety. Follow the recommended safety guidelines provided in the alert, such as seeking shelter, avoiding outdoor activities, or evacuating if necessary.

3. Stay Informed: Continuously monitor weather alerts on your Baofeng radio to stay informed about changing weather conditions in your area. Remain vigilant and prepared to respond to evolving weather patterns, keeping yourself and your community safe from harm.

4. Empower Yourself: By activating weather alert features on your Baofeng radio, you empower yourself with timely and critical information about weather-related threats, enabling you to make informed decisions and take proactive measures to protect yourself and your loved ones.

5. Stay Safe and Informed: With your Baofeng radio as your weather alert companion, you can stay safe and informed, even in the midst of severe weather events. Take advantage of its powerful capabilities to stay ahead of the storm and ensure your well-being in any weather condition.

Activating this functionality on your Baofeng radio is incredibly simple. Just follow the steps mentioned below and you'll know what to do when the weather isn't on your side.

1. **Enter Weather Alert Menu**: Navigate to the menu settings on your Baofeng radio and locate the weather alert feature. Activate weather alert mode to enable the radio to scan for and receive NOAA weather alerts.

2. **Select Alert Types**: Choose the types of weather alerts you wish to receive, including tornado warnings, severe thunderstorm warnings, flash flood warnings, and more. Customize your alert settings to suit your specific location and weather-related concerns.

3. **Set Alert Tone**: Configure the alert tone settings on your Baofeng radio to ensure that you're alerted promptly when a weather alert is received. Choose a distinct and recognizable tone that will grab your attention, even in noisy environments.

Emergency Communication Plans

Up till now, we've covered that you use your Baofeng radio to communicate on emergency channels and frequencies and can configure it to receive prompt weather alerts. But let's face the reality here, and that reality is that all of this information will not be of any good unless you turn it into an actionable plan.

All of us have heard the quote *"If you fail to plan, you plan to fail,"* right? Well, this quote holds true, but I'm here to help you make it something you don't have to experience in your life. At least not when you have the Baofeng radio on your side.

By creating a communication plan you'll always have access to emergency services and will be able to coordinate you resources and share information effectively. However, before you develop a communication plan, you need to determir what your communication objectives are. This can include things like:

- Setting up communication channels.
- Choosing emergency contact points.
- Making sure emergency information can be sent swiftly.

Having clearly defined objectives will give you more focus and clarity when you're developing a communication plar Once you've defined your objectives, you can follow the steps below:

Creating Contact Lists and Channels

a. **Compile Contact Information**: Gather contact details of essential individuals, including famil members, neighbors, emergency services, and community leaders. Create a comprehensive contact lis to ensure you can reach out to key stakeholders when needed.

b. **Assign Communication Channels**: Allocate specific communication channels for different scenario such as using Baofeng radio frequencies for local communication, mobile phones for long-distanc communication, and alternative methods like hand signals or flares for emergency signaling.

2. **Establishing Communication Protocols**

a. **Define Communication Protocols**: Establish clear protocols for initiating communication, relayin messages, and confirming receipt of information. Standardized procedures streamline communicatio processes, minimizing confusion and enhancing efficiency during emergencies.

b. **Practice Effective Radio Etiquette**: Familiarize yourself with basic radio etiquette principles, such a using clear and concise language, speaking slowly and clearly, and using standardized phrases like "over and "out" to indicate the end of transmissions.

3. **Conducting Regular Drills and Exercises**

a. **Schedule Communication Drills**: Conduct regular communication drills and exercises to test th effectiveness of your communication plans and identify areas for improvement. Simulate differen emergency scenarios to assess your readiness and refine your response strategies accordingly.

b. **Evaluate and Adjust Plans**: After each drill, evaluate the performance of your communication plans an make necessary adjustments based on feedback and lessons learned. Continuously refine your plans t ensure they remain relevant and effective in evolving circumstances.

4. **Maintaining Redundant Communication Options**

a. **Diversify Communication Methods**: Implement redundant communication options to mitigate the ris of communication failure during emergencies. In addition to Baofeng radios, consider alternativ communication devices such as satellite phones, CB radios, and signal mirrors.

b. **Regular Equipment Maintenance**: Ensure your communication equipment, including Baofeng radios, i well-maintained and operational at all times. Conduct regular checks, replace batteries as needed, anc carry spare parts to address any technical issues that may arise.

Search and Rescue Missions

You can also use your Baofeng radio to carry out search and rescue missions. The Baofeng radio can help in navigating terrains and establishing communication protocols. By using these capabilities, you can easily locate and rescue missing persons, stranded bikers, or individuals in distress.

However, before you carry out such missions, there is a set of instructions that you need to follow for each aspect of the search and rescue missions. These aspects along with their instructions are mentioned below.

1. **Navigating Terrain with Baofeng Radios**

    a. **Set Up Radio Frequencies**: Before embarking on a search mission, program your Baofeng radio with preset frequencies designated for search and rescue operations. Ensure all team members use the same frequencies to facilitate seamless communication.

    b. **Utilize VFO Mode for Flexibility**: In Variable Frequency Operation (VFO) mode, use your Baofeng radio to scan for available frequencies and identify potential communication channels with other search teams or base stations.

2. **Effective Communication Strategies**

    a. **Establish Communication Protocols**: Define clear communication protocols for relaying information, reporting findings, and coordinating search efforts. Standardized procedures minimize confusion and enhance efficiency during high-pressure situations.

    b. **Employ Dual Watch Feature**: Activate the Dual Watch feature on your Baofeng radio to monitor two frequencies simultaneously, enabling you to listen for distress calls while maintaining communication with your team members.

3. **Utilizing Location Beacons and Signaling Devices**

    a. **Deploy Location Beacons**: Equip search team members with location beacons or GPS devices to facilitate precise navigation and track their movements in real-time. Baofeng radios can communicate with these devices to relay location information and coordinate search efforts effectively.

    b. **Employ Visual Signaling Techniques**: In addition to radio communication, utilize visual signaling techniques such as hand signals, signal mirrors, or smoke signals to attract attention and communicate with individuals in remote or obstructed areas.

Executing Search and Rescue Protocols

    c. **Conduct Grid Searches**: Organize search teams into designated grid patterns to systematically cover the search area and maximize coverage. Maintain regular communication with team members to track progress and coordinate search efforts effectively.

    d. **Responding to Distress Signals**: Upon receiving distress signals or locating individuals in need of assistance, communicate their location and condition to the appropriate authorities or rescue teams for prompt response and evacuation.

CHAPTER 8:
Ham Radio Operations

By now, you've mastered using the Baofeng radio in emergency situations, are familiar with advanced functionalities, signals tones and more. In this chapter, we'll cover ham radio operations. We'll cover the basics, the frequencies, learn about obtaining a license for a ham radio, and more. So, let's get started.

Ham Radio Practices

Ham radio operators often engage in recreational communication,

emergency preparedness, and technical experiments and believe it or not, but it's a fulfilling hobby for people of all ages. Now, when it comes to covering ham radio practices, there are quite a few things that you'll need to learn. Let's go over each on them:

1. **Exploring Ham Radio Basics**

    a. **Obtain a Ham Radio License**: Before transmitting on amateur radio frequencies, you need to obtain a ham radio license from your country's regulatory authority. Doing so ensures that operators understand the procedures, regulations, and technical principles essential for safe and responsible radio operation.

    b. **Learn Operating Procedures**: Familiarize yourself with standard operating procedures, including call signs, Q codes, and phonetic alphabets used for clear and concise communication. When you adhere to established protocols, it can enhance communication efficiency and minimize misunderstandings during radio transmissions.

2. **Frequency Allocation and Band Usage**

   a. **Understand Frequency Bands**: Ham radio operates across a wide range of frequency bands allocated by regulatory authorities for amateur use. Knowing the band plans and frequency allocations can help you ensure compliance with legal and operational requirements.

   b. **Explore Band Conditions**: Monitor band conditions and propagation characteristics to identify optimal frequencies for communication. Use tools such as propagation prediction software and online resources to anticipate band openings and maximize contact opportunities.

3. **Participating in Ham Radio Nets**

   a. **Join Local Nets**: Engage with local amateur radio nets, join gatherings of operators who meet regularly to exchange information, discuss topics of interest, and provide support during emergencies. Participating in nets fosters community involvement and enhances your radio operating skills.

   b. **Conduct Net Operations**: Follow net protocols and guidelines established by net controllers to contribute effectively to net discussions. Listen attentively, wait for your turn to transmit, and adhere to net procedures to maintain orderly communication flow.

4. **Embracing Ham Radio Culture**

   a. **Attend Ham Radio Events**: Immerse yourself in the vibrant ham radio community by attending club meetings, field days, and radio rallies. These events offer opportunities to learn from experienced operators, exchange ideas, and showcase your radio projects.

   b. **Engage in Technical Projects**: Explore the technical aspects of amateur radio by undertaking projects such as antenna construction, radio modifications, and digital mode experimentation. Hands-on projects enhance your understanding of radio technology and foster creativity in radio operation.

Obtaining a Ham Radio License

Next up is licensing. Before you operate a ham radio you need to have a license. This ensures that you have the skills and expertise needed to operate the equipment safely and with responsibility. It also ensures that you're adhering to legal regulations and frequency allocations.

Once you have your license you'll become a full-fledged member of the amateur radio community. However, to get your license, you need to follow the steps for each stage of the process mentioned below:

1. **Preparation and Study Materials**

   a. **Acquire Study Resources**: Begin with acquiring study materials tailored to the licensing exam for your country. These resources may include study guides, practice exams, and online courses designed to help you master the fundamental concepts of radio theory, regulations, and operating procedures.

   b. **Join Licensing Classes or Study Groups**: Enroll in licensing classes or join study groups led by experienced ham radio operators. Collaborating with fellow enthusiasts provides valuable support, guidance, and encouragement throughout the learning process, increasing your chances of success on the exam.

2. **Taking the Licensing Exam**

    a. **Register for the Exam**: Locate an authorized testing center in your area and register for the ham radio licensing exam. Be sure to check the exam requirements, including identification and fees, and arrive prepared on exam day with the necessary materials.

    b. **Prepare for Success**: Dedicate ample time to review study materials, practice exam questions, and reinforce your understanding of key concepts. Focus on areas where you may need additional study and approach the exam with confidence in your abilities.

3. **Passing the Exam and Obtaining Your License**

    a. **Ace the Exam**: On exam day, approach each question methodically, carefully reading and understanding each prompt before selecting your answers. Trust in your preparation and remain calm and focused throughout the exam to maximize your chances of success.

    b. **Receive Your License**: Upon passing the exam, celebrate your achievement as you await the arrival of your ham radio license from the regulatory authority. Your license grants you access to amateur radio frequencies and privileges, allowing you to begin transmitting and communicating with fellow operators.

Ham Bands

Once you're a part of the community, the next thing you need to do is familiarize yourself with is the Ham bands are basically High-Frequency (HF) bands used for long distance communication. Ham bands also include Very High Frequency/Ultra High Frequency (VHF/UHF) bands.

These are the ones that are used for both local and regional coverage. These bands will give you a lot of different opportunities to explore. But, if you do that, some of the key things that you need to be aware of include:

Exploring Frequency Allocation

    a. **Research Frequency Allocations**: Visit the website of your country's telecommunications regulatory authority to find a list of allocated amateur radio bands.

    b. **Program Baofeng for Ham Bands**: Turn on your Baofeng radio and press the "Menu" button. Press "Menu" once more after using the arrow keys to select the "Frequency Mode" option. Utilizing the numeric keypad, enter the preferred frequency. Press the "Menu" button again to save the frequency.

Optimizing Operating Parameters

    c. **Adjust Transmit Power and Modulation**: Press the "Menu" button and navigate to the "TX Power" option. Select the desired power level (e.g., Low/Medium/High) using the arrow keys, then press "Menu" to confirm. For modulation, press the "Menu" button and navigate to the "Modulation" option. Choose the modulation type (e.g., FM/NFM) and press "Menu" to confirm.

    d. **Configure Repeater Offset and Tone**: Press the "Menu" button and navigate to the "Repeater Shift" option. Use the arrow keys to select the appropriate offset value (e.g., 600 kHz for 2m band) and press "Menu" to confirm. To configure CTCSS/DCS tones, navigate to the "Tone" option and select the desired tone frequency (e.g., 100.0 Hz). Press "Menu" to save the settings.

ngaging in Ham Radio Activities

  e. **Participate in Contests and Events**: Join local ham radio clubs or online communities to stay updated on upcoming contests and events. Tune your Baofeng radio to the specified contest frequencies and listen for activity. When ready to transmit, announce your callsign and make your contacts.

  f. **Conduct DX (Long-Distance) Communication**: Experiment with different HF bands and antenna configurations to optimize your chances of making DX contacts. Monitor propagation conditions using online tools or mobile apps, and aim your antenna towards the desired region for optimal signal strength.

ngaging in Ham Radio Nets

adio nets are basically organized gatherings of amateur radio operators who meet regularly and talk about different opics they're interested in. This includes things like emergency communication protocols, new features and nctionalities, and more.

y taking part in these nets, you can amp up your skills, share what you know with others, and develop relationships ith like-minded individuals. Sounds pretty cool, doesn't it? Who knows, maybe we even get to meet each other in one f them. So, with that in mind, here's how you can find, join, and participate in ham radio nets.

1. Finding and Joining Ham Radio Nets

  a. **Research Local Nets**: Explore online resources, forums, or social media groups to identify ham radio nets in your area. Look for nets that align with your interests and operating capabilities, such as general discussion nets, emergency preparedness nets, or technical discussion nets.

  b. **Tune Baofeng to Net Frequency**: Turn on your Baofeng radio and press the "VFO/MR" button to enter VFO mode. Use the arrow keys to tune to the frequency of the desired net, typically announced in advance by net organizers. Press the "Menu" button, select "Frequency Mode," and enter the net frequency using the numeric keypad. Press "Menu" to save the frequency.

2. Participating in Ham Radio Nets

  a. **Listen First**: Upon tuning to the net frequency, listen carefully to ongoing conversations to familiarize yourself with the net's format, rules, and protocols. Take note of the net control station's call sign and any instructions provided.

  b. **Check In**: When invited by the net control station, key your Baofeng radio and announce your callsign to check in to the net. Follow any specific check-in procedures specified by the net control, such as stating your location or providing signal reports.

  c. **Contribute to Discussions**: Engage actively in net discussions by sharing relevant information, asking questions, or offering assistance to other participants. Respect net etiquette, wait for your turn to speak, and avoid interrupting ongoing.

CHAPTER 9:
Weather Monitoring and NOAA Frequencies

Planning a trip to a remote island or location where you'll camp, hunt, and hike trails?

Well, you'll need quite a lot of gear for that and the most important item will be your Baofeng radio. Think about it for a second: You're out in the wild on a hunting run and all of sudden the sky turns gray and out comes pouring rain. It's the worst thunderstorm you could possibly imagine.

The terrain below your feet, that was solid a few minutes ago, is now

all muddy and slippery. Your chances of moving around like a ninja are very thin! As you continue to hunt, it become easier for you to find prey and easier to become one given that you might fall down and be caught off guard.

But all of this can be avoided if you know how to use the NOAA frequencies on your Baofeng radio and that's exactl what we're going to learn about in this chapter. So, with hopes of not being a prey, let's begin!

Accessing Critical NOAA Weather Frequencies

These frequencies are what the National Oceanic and Atmospheric Administration (NOAA) uses to broadcast weathe conditions. They are highly essential when it comes to communication during severe weather conditions.

y tuning into these frequencies, you have access to the latest weather conditions and forecasts, warnings, and emergencies. You can then use this information to plan proactively to avoid challenges that you might face otherwise.

o access and use the Critical NOAA Weather Frequencies on your Baofeng radio all you have to do is follow these eps below.

ccessing NOAA Weather Frequencies

- **Enter Frequency**: Use the radio's keypad to enter the frequency of your local NOAA Weather Radio station. NOAA broadcasts typically operate within the VHF frequency range, commonly around 162.400 to 162.550 MHz.
- **Select Channel**: Once you've entered the frequency, press the appropriate button or menu option to save it as a channel for quick access.
- **Enable NOAA Mode**: Switch the radio to NOAA mode to ensure optimal reception of weather broadcasts.
- **Monitor NOAA Broadcasts**: Tune into the designated NOAA channel and adjust the volume to a comfortable level. You'll now be able to receive weather updates and alerts as they are broadcasted by NOAA.

perating on NOAA Weather Frequencies

- **Scan Function**: Activate the scan function on your Baofeng radio to continuously monitor NOAA frequencies for incoming weather broadcasts.
- **Weather Alert Feature**: Enable the weather alert feature on your radio to receive automatic alerts when NOAA issues severe weather warnings for your area.
- **Manual Tuning**: Use manual tuning capabilities to fine-tune the radio's frequency and optimize reception of NOAA broadcasts, especially in areas with weak signal strength.

Configuring Weather Channels

Now that you're aware of how you can operate the weather frequencies, let's learn about configuring weather channels n your Baofeng radio. Doing so will help you customize and organize NOAA Weather Radio frequencies for quick and asy access.

'ou'll be able to streamline the process for monitoring weather broadcasts and be more prepared than ever before. To onfigure and operate weather channels, here's what you need to do.

1. **Configuring Weather Channels**

    a. **Access Menu**: Press the "Menu" button on your Baofeng radio to enter the menu mode.

    b. **Select Channel Mode**: Navigate through the menu options using the arrow keys and select the channel mode setting.

    c. **Enter Weather Frequencies**: Use the keypad to enter the frequencies of your local NOAA Weather Radio stations. Refer to reliable sources or NOAA's official website for accurate frequency information.

    d. **Save Channels**: Once you've entered the frequencies, save them as weather channels by selecting the corresponding channel number and pressing the "Save" or "Confirm" button.

e. **Organize Channels**: Arrange the weather channels in sequential order for easy access. You can reorder channels using the channel editing function in the radio's menu.

f. **Exit Menu**: After configuring and organizing weather channels, exit the menu mode to return to normal radio operation.

2. **Operating Weather Channels**

a. **Channel Selection**: Use the channel selector knob or keypad to navigate between configured weather channels.

b. **Scan Function**: Activate the scan function to automatically scan through weather channels for incoming broadcasts.

c. **Channel Lock**: Lock weather channels to prevent accidental changes or deletions, ensuring consistent access to critical weather information.

Severe Weather Alerting

Another great thing about the Baofeng radio is its severe weather alert capabilities. By activating these features, you can receive automatic notifications about weather emergencies. Once you have this feature activated, you'll automatically stay one-step ahead of dangerous situations. It'll give you real-time updates which you can use to make informed decisions. To activate the alerts, here's what you need to do:

1. **Access Menu**: Press the "Menu" button on your Baofeng radio to enter the menu mode.

2. **Select Alert Setting**: Navigate through the menu options using the arrow keys and select the "Alert" or "Weather Alert" setting.

3. **Enable Alert**: Toggle the alert function by selecting "On" or "Enable." This activates the radio's ability to receive weather alerts.

4. **Set Alert Mode**: Choose the desired alert mode, such as "Tone" or "Voice," to determine how alerts are signaled.

5. **Save Settings**: Once configured, save the settings by pressing the "Save" or "Confirm" button.

Receiving Severe Weather Alerts

Once the Severe Weather Alert function is activated, your Baofeng radio will automatically monitor NOAA Weather Radio broadcasts for severe weather alerts. When an alert is issued for your area, the radio will emit a tone or voice alert to notify you of how things might turn out. .

APRS Integration

The automatic packet reporting system (APRS) is a digital communication protocol that's used by amateur radio operators when they are transmitting and receiving location information, reports about the weather, or other messages. To setup this protocol and operate it, here's what you need to do:

1. **Setting Up Baofeng for APRS**

a. **Enter APRS Frequency**: Press the "VFO/MR" button on your Baofeng radio to enter VFO mode. Tune to the APRS frequency, typically around 144.390 MHz in the United States, using the arrow keys. Press the "Menu" button, navigate to the "Frequency Mode," and input the APRS frequency using the numeric keypad. Press "Menu" to save the frequency.

b. **Configure TNC Settings**: Access the Baofeng radio's menu and navigate to the TNC (Terminal Node Controller) settings. Ensure that the TNC is enabled and set to the appropriate baud rate for APRS communication, typically 1200 baud.

c. **Connect to GPS Device**: If using a GPS-enabled Baofeng model, connect your GPS device to the radio using the appropriate cable. Ensure that the GPS data output is configured to interface with the Baofeng radio.

2. **Operating APRS with Baofeng**

a. **Transmit Location Beacons**: Once configured, your Baofeng radio will automatically transmit location beacons at regular intervals, typically every few minutes. These beacons contain GPS coordinates, altitude, speed, and other relevant information.

b. **Send APRS Messages**: Use the messaging functionality of APRS to send short text messages to other APRS-equipped stations. Press the "Menu" button, navigate to the messaging menu, and enter the callsign of the recipient station along with your message. Press "Send" to transmit the message.

c. **Track Nearby Stations**: Monitor the APRS network to track the positions of nearby stations and objects. Use the Baofeng radio's display to view incoming APRS packets, including station callsigns, locations, and other telemetry data.

CHAPTER 10:
Satellite Communication

～❧❧❧❧～

The Baofeng radio is a truly unique asset that can help you communicate even when traditional methods are not available. How? Allow me to explain. Let's go back to our example from earlier. You are the one where you took a trip to the remote island. Remember that?

Alright, so now, let's say that traditional communication methods are down and you have to reach someone not on the island. What would you do in such a scenario? Well, if you have your Baofeng radio with you don't have to rely on the traditional methods. You can use satellite communication.

Yes, you read that right! How are you going to do that? Well, that's what we're going to cover in this chapter.

Satellite Communication

Having the ability to communicate via satellite signal opens up so many more possibilities for you to connect with others. The way this works is a bit technical, but you and I, we're going to keep it simple. So, with satellite communication, what happens is that your Baofeng radio transmits a signal to an orbiting satellite.

The satellite then acts as your relay station which basically allows you to communicate over long distances. Knowing how to use your Baofeng radio for satellite communication can benefit you regardless of whether you're an operator, emergency responder, or someone going on an adventure.

To access satellite frequencies, here's what you need to do:

1. **Enable Frequency Mode**: Press the "VFO/MR" button on your Baofeng radio to enter frequency mode.

2. **Select Satellite Frequency**: Use the arrow keys to tune to the desired satellite frequency. Satellite frequencies are typically listed in satellite tracking software or online databases.

3. **Adjust Settings**: Ensure that your radio is set to the appropriate transmit power, modulation mode, and other settings required for satellite communication.

4. **Antenna Orientation**: Orient your antenna for optimal signal reception. Satellite tracking software or online resources can provide guidance on the satellite's position and optimal antenna orientation.

5. **Transmit and Receive**: When the satellite passes overhead, transmit your message using the push-to-talk (PTT) button on your Baofeng radio. Listen for responses from other users or ground stations.

Now that you know how to access satellite frequencies, take a look at these key tips to make the most of the communication and have fun!

- **Timing**: Plan your communication sessions to coincide with satellite passes for optimal signal strength.

- **Antenna Considerations**: Invest in a high-gain directional antenna for improved satellite communication performance.

- **Practice**: Familiarize yourself with satellite tracking software and practice communicating with other users to improve your skills.

Capturing Satellite Passes

The next thing we're going to cover is satellite passes. You see, satellite passes are the predictable paths in which satellites orbit the Earth. You can track and predict these paths using software and online tools. Once you've predicted these paths, all you have to do is align your antenna to the path as the satellite moves across the sky.

This will allow you to use the satellite and communicate over long distances. To prepare for the passes and capture them, here's what you need to do:

1. **Preparing for Satellite Passes**

    a. **Satellite Tracking Software**: Download and install satellite tracking software on your computer or smartphone. Popular options include Orbitron, GPredict, and Heavens Above.

    b. **Antenna**: Use a high-gain directional antenna for optimal signal reception. Position the antenna in an open area with a clear view of the sky.

    c. **Frequency Information**: Obtain frequency information for the satellite you wish to track. This can be found in satellite tracking software or online databases.

2. **Capturing Satellite Passes**

    a. **Enable Frequency Mode**: Press the "VFO/MR" button on your Baofeng radio to enter frequency mode.

    b. **Select Satellite Frequency**: Use the arrow keys to tune to the frequency of the satellite you wish to track.

    c. **Align Antenna**: Orient your antenna according to the predicted path of the satellite. Satellite tracking software will provide real-time information on the satellite's position.

    d. **Monitor Pass**: As the satellite passes overhead, monitor your radio for signals. Adjust the antenna position if necessary to maintain a strong signal.

e. **Transmit and Receive**: When the satellite is within range, use the push-to-talk (PTT) button on you Baofeng radio to transmit your message. Listen for responses from other users or ground stations.

Fine-Tuning for Doppler Shifts

Ever heard of doppler shifts? No? You see a Doppler shift basically occurs when a satellite is drawn near or and as it goe aways from Earth.

As a satellite comes closer to the Earth, the transmission frequencies get elevated and as it goes away, they reced Now, I know that this is something that's out of your control. In most cases when something is out of control, yo shouldn't think about it too much.

But that's not the case here. If you ignore these shifts, your transmission signals can get distorted. So, you communication efforts might be as effective as you need them to be and that's not a good thing. However, you can us these shifts to your benefit and communicate easily. To do that:

1. **Calculate the Shift**: Compute the anticipated Doppler shift for your satellite rendezvous, leveraging online too or satellite tracking software to unveil its magnitude.

2. **Adjust with Precision**: Harnessing your newfound knowledge, enact precise adjustments to the transm frequency on your radio, aligning it with the calculated Doppler shift.

3. **Maintain Vigilance**: Amidst communication sessions, maintain a vigilant watch over signal strength and qualit paying attention to any fluctuations indicative of Doppler variations.

4. **Adapt in Real-Time**: As the satellite meanders through its celestial path, adapt dynamically, tweaking transm frequencies in real-time to harmonize with Doppler oscillations.

5. **Optimize Communication**: Through meticulous fine-tuning, unlock the full potential of satellit communication, securing a seamless exchange of signals and augmenting the breadth of your communicativ prowess.

Satellite Communication Challenges

I understand that satellite communication is something that's really exciting. While using your Baofeng radio t communicate via satellite has a unique feel to it, there are some challenges here that you need to be aware of. Some o these challenges include signal interference and orbital dynamics.

These things might seem a bit difficult to understand and to be honest they are. They were for me, at least. But what want you to know is that learning about these challenges and how to tackle them is very important. Once you'v mastered this, you'll be able unlock the full potential of your Baofeng radio. Now, isn't that an exciting goal to chase?

So, with that in mind here are some common challenges and solutions that you need to be aware of.

Common Challenges

- **Signal Interference**: Radio frequency interference from nearby electronic devices or natural phenomena ca disrupt satellite communication. This interference may manifest as static or distortion in received signals.

- **Orbital Dynamics**: Satellites orbit the Earth at high speeds and altitudes, resulting in rapid changes in their positions relative to ground stations. Tracking and maintaining communication with moving satellites can be challenging.

- **Antenna Alignment**: Achieving precise antenna alignment is crucial for capturing satellite signals effectively. Misalignment or obstructions can lead to weak or lost signals.

ractical Solutions

- **Minimize Interference**: Position your Baofeng radio away from sources of electromagnetic interference, such as computers, routers, and power lines. Additionally, consider using a ferrite choke or shielded cables to reduce interference.

- **Utilize Tracking Software**: Leverage satellite tracking software to predict the positions of satellites in real-time. This allows you to anticipate satellite passes and adjust your antenna accordingly for optimal signal reception.

- **Optimize Antenna Alignment**: Invest time in calibrating and aligning your antenna for maximum signal strength. Use a compass and inclinometer to ensure accurate azimuth and elevation adjustments, and periodically recheck alignment during extended communication sessions.

CHAPTER 11:
Baofeng Antenna Projects

Raising your eyebrows after reading the title? Well, keep them raised, because in this chapter we're going to learn how you can whip up your own antenna! Yeah, that's right.

Why would you need to do that? Let's go back to our camping trip example again. This time as you continue on your trek the path is a bit challenging and slippery. Because of this you wobble, but you keep yourself from falling down. You're in the clear. But your Baofeng radio is not!

Don't get me wrong, the Baofeng radio is a solid and durable little device that'll last

for decades. But it's not immune to accidents. And an accident is exactly what happens when you wobble. In that split second, the antenna on your Baofeng radio hit the side of a rock, and now, it's RIP for the antenna.

Despite this you managed to make it back home, but now you have a radio without an antenna. Let's look at what we can do about that!

The Pivotal Role of the Antenna and Replacement Options

Before we dive into what you can do about the antenna, you need to understand its role first. It's responsible for sending and receiving radio waves and impacts how far the signals reach and how clear they are.

So, it basically controls the effectiveness of your communication efforts. You may feel the need to replace the antenna because the current range is not suitable for you, the current one broke, or managed to save up some cash and feel like it's time for an upgrade. In either case, some of the options you have include:

1. **Aftermarket Antennas:** Upgrading to a high-quality aftermarket antenna is one of the simplest yet most impactful modifications you can make. Aftermarket antennas are specifically designed to enhance signal transmission and reception over greater distances and through challenging terrains, a necessity for guerrilla units needing to maintain communication over dispersed areas.

2. **SMA to BNC Adapter Conversion:** The Baofeng radio, like many of its counterparts, comes equipped with an SMA (SubMiniature version A) connector for the antenna. While functional, SMA connectors are not known for their durability or ease of use in the field. Converting the SMA connection to a BNC (Bayonet Neill-Concelman) connector offers several advantages:

   a. **Durability:** BNC connectors are more robust and can withstand the rigors of field operations better than their SMA counterparts. Their design reduces the risk of damage during rapid deployment or in harsh conditions.

   b. **Ease of Use:** BNC connectors allow for quick and tool-free attachment and detachment of antennas. This feature is invaluable in situations where time is of the essence or when conditions necessitate rapid changes in equipment configuration.

   c. **Versatility:** By converting to a BNC connector, operators gain access to a broader range of antennas, including those designed for specific operational needs or those improvised in the field. This versatility can be a critical advantage in adapting to the evolving demands of guerrilla warfare.

Practical Instructions for Conversion and Antenna Upgrade

Now that you've gone off the different option, you probably have an idea of what it is you're going to do. So, let's say that you've found the perfect antenna for your Baofeng radio. To attach it to the radio, here's what you're going to do:

1. **Acquire a Quality SMA to BNC Adapter:** These adapters are readily available from electronic parts suppliers and online. Ensure the adapter is compatible with your Baofeng model.

2. **Attach the Adapter:** Carefully screw the SMA end of the adapter into the antenna port of the Baofeng radio. Avoid over tightening to prevent damage.

3. **Select and Attach an Aftermarket Antenna:** Choose an antenna that meets your operational needs—consider factors such as frequency range, gain, and size. Attach the antenna to the BNC connector, ensuring a secure fit.

4. **Test Your Setup:** Before deploying into the field, test the new antenna setup in various conditions to familiarize yourself with its performance characteristics and to ensure it meets your communication requirements.

Radio Theory Basics for Effective Communication

The next thing we're going to cover is some basic theories about radio communication. To do this, we'll look at how you can maximize the communication range and how the environment impacts radio communication. I'll even share some tips with you on how to optimize communication in different environments.

However, before we get into all of that you need to know that the Baofeng radio is basically a line-of-sight device. What does that mean? Well, to keep it simple, this means that things like obstacles and the overall environment can, indeed, influence your Baofeng radio's ability to send and receive signals. As far as line of sight and communication are concerned:

- **Terrain:** VHF and UHF radio waves travel in straight lines and cannot bend around the Earth's curvature. Hence the communication range in flat areas is generally better than in hilly or mountainous terrain. However, VHF waves can diffract over hills to some extent, making them slightly better in uneven terrain than UHF signals.

- **Obstructions:** Physical obstacles, such as buildings in urban settings or dense forests, can block or weaken VHF and UHF signals. UHF frequencies, due to their shorter wavelength, are better at penetrating through dense materials and between buildings.

- **Line-of-Sight:** To maximize the communication range, ensure a clear line of sight between the transmitting and receiving radios. The higher the antenna's placement, the longer the line of sight and, consequently, the greater the potential communication distance.

Remember when I said that the antenna is more important than you think? Great! It can also help you tackle any link of sight communication issues because it can be used to maximize the communication range. To do that, the three things that you need to consider include:

- **Antenna Type:** The choice of antenna significantly affects a radio's performance. While the stock antenna that comes with most handheld radios is adequate for close-range communication, upgrading to a higher-quality aftermarket antenna can dramatically improve both range and signal clarity.

- **Antenna Placement:** Optimal antenna placement can help overcome environmental limitations. For field operations, consider using a makeshift elevated antenna or relocating to higher ground to extend your radio reach.

- **Antenna Length:** The antenna's length should ideally match the wavelength of the frequency being used. A resonant antenna, tuned to the specific frequency of operation, will be more efficient at transmitting and receiving signals.

Environmental Impact on Radio Communications

When it comes to the environment and radio communication, you need to know that the environment can influence how the radio waves are transmitted and how well your radio performs. However, this influence can vary among the different types of environments. With that in mind, let's look at how this plays out and what you can do to optimize communication.

Urban Environments

- **Signal Reflection and Multipath:** In urban areas, dense concentrations of buildings and other structures can cause radio waves to reflect and scatter, leading to a phenomenon known as multipath interference. This can either weaken the signal or cause it to arrive at the receiver from multiple paths, potentially causing confusion and distortion.

- **Penetration Losses:** Concrete and metal structures can reduce the strength of radio signals, especially UHF frequencies. While UHF is generally better at penetrating urban obstructions than VHF, significant losses can occur inside buildings or underground.

Practical Tips:

- Utilize UHF frequencies for better penetration in urban settings.

- Consider the placement of repeaters on high buildings to enhance signal coverage.
- Experiment with directional antennas to focus signals and mitigate multipath interference.

Rural Environments

- **Terrain Influence:** The relatively open spaces of rural environments allow for longer line-of-sight distances, but terrain features like hills and valleys can obstruct signals. VHF frequencies, which can diffract over obstacles to a degree, are generally more effective in these settings.
- **Vegetation Losses:** Dense foliage can absorb radio signals, particularly at higher frequencies. The impact is more pronounced with UHF than with VHF.

Practical Tips:

- Opt for VHF frequencies for broader coverage in hilly or forested rural areas.
- Elevate antennas as much as possible to extend line-of-sight across terrain obstacles.
- Use omni-directional antennas for general area coverage, switching to directional antennas for specific, distant communication needs.

Jungle Environments

- **High Attenuation:** The dense vegetation of jungle environments can severely attenuate radio signals, reducing range and clarity. This effect is more significant for UHF signals due to their shorter wavelengths.
- **Limited Line-of-Sight:** The thick canopy and undergrowth limit line-of-sight, making high placement of antennas challenging but crucial for effective communication.

Practical Tips:

- Prioritize VHF frequencies for their better performance in penetrating foliage.
- Deploy wire antennas vertically to increase the effective height and improve signal propagation.
- Regularly test communication paths and be prepared to adjust frequency use and antenna configurations as environmental conditions change.

Alright, so now that we've covered some theoretical knowledge, it's time to put all of that into action. To effectively maxout on your Baofeng radio's communication range, you need to follow the steps mentioned below:

1. Evaluate Your Environment: Before deploying your radios, assess the terrain and identify potential line-of-sight obstructions. Plan your antenna placement and communication points accordingly.
2. Opt for High-Ground: Whenever possible, operate from elevated positions or use elevated antenna setups to enhance line-of-sight and extend your communication range.
3. Upgrade Your Antenna: Consider investing in higher-quality, band-specific antennas for your radios. A simple upgrade from the standard antenna to a more efficient model can result in significantly better performance.
4. Test and Adapt: Regularly test your radios in different environments and with different antenna setups to understand their capabilities and limitations. Adapt your communication strategy based on these insights for optimal results.

Antenna Theory

I know this chapter is a bit long, but stick with me here, because all of this is essential for effective communication. Believe me, it'll come in handy when you have limited resources and aren't sure what the environment will be like.

So, as we get started with antenna theory, the first thing you need to know about is Resonance. This is when the antenna's length is perfectly matched to the frequency's wavelength it needs to transmit. I know you're scratching your head right now and are probably thinking about what this means. I had a similar reaction when I first heard about this.

Now, to figure what the length should be you need to do some calculations. We'll get to that in a second, but first, let's look at resonant length and wavelength calculation:

1. **Resonant Length:** A resonant antenna is tuned to a specific frequency, meaning its physical length is a certain fraction (commonly 1/4 or 1/2) of the wavelength of the radio waves it is designed to handle. This tuning optimizes the antenna's ability to radiate or receive radio waves efficiently, leading to clearer communication and extended range.

2. **Wavelength Calculation:** The wavelength (λ) of a radio wave in meters can be calculated by dividing the speed of light (c, approximately 300,000,000 meters per second) by the frequency (f) in hertz. For practical antenna construction, it's crucial to know the wavelength to determine the optimal antenna length for a given frequency.

Determining Antenna Length and Steps For Construction

If you want to create an antenna that functions effectively, you need to know how to calculate the length. I know that calculations can get a bit tricky, so let's take little baby steps towards our end goal. The first thing you need to know is that the basic principle of these calculations is tied to the wavelength (λ) that you're trying to transmit or receive.

In addition, as the frequency increases, the wavelength decreases, and vice versa. The most common type of antenna is the half-wave dipole. The length of the antenna should be half of the wavelength of the target frequency. The formula to calculate the length of a half-wave dipole antenna (L) in meters is:

$$L\text{(in meters)} = \frac{150}{f\text{(in MHz)}}$$

This formula gives the total length of the dipole from one end to the other. Suppose you want to build a dipole antenna for the 2-meter amateur radio band, centered around 146 MHz. Using the formula:

1. Calculate the wavelength:

   $$\lambda = \frac{300}{146} \approx 2.05 \text{ meters}$$

2. Calculate the length of a half-wave dipole:

   $$L = \frac{150}{146} \approx 1.03 \text{ meters}$$

So, what this basically means is that each leg of the dipole should be approximately half of 1.03 meters, or about 51.5 cm long. Now to construct such an antenna, there are a few things you need to keep in mind. This includes:

1. Material Selection: Antennas can be made from a wide variety of conductive materials, but in field conditions, simplicity and availability often dictate choices. Copper wire is a popular choice due to its excellent conductivity.

and flexibility. However, in an austere environment, even materials like metal tape measures or salvaged wire can be repurposed into effective antennas.

2. Improvisation in the Field: In situations where precise construction is not possible, trial and error, coupled with field testing, can lead to a satisfactorily resonant antenna. Adjustments in antenna length can be made by incrementally trimming or extending the antenna and testing performance until the optimal length is found.

3. Enhancing Performance with Antenna Tuners: When exact resonance cannot be achieved due to operational constraints, an antenna tuner can be used to electronically adjust the antenna's resonance. This device can make a non-resonant antenna work acceptably well over a range of frequencies, providing flexibility in field operations.

Antenna Construction

Now that you know how you can make the most basic type of antenna, let's shift our focus to how you can improvise the construction using common materials. Two of the most important things you'll need to focus on include:

- **Insulators:** These are used at the ends of the antenna elements and at any point where the antenna wire must be supported or separated from conductive materials (such as metal supports or the earth). Insulators prevent unwanted electrical contact that can detune the antenna or cause energy loss. Common insulator materials include plastic, wood, or even dry air space for a temporary setup.

- **Element Measurement:** The length of the antenna elements is paramount in defining the antenna's resonant frequency – the frequency at which the antenna most efficiently radiates or receives radio waves. Incorrect measurements can lead to poor performance, including weak signal transmission and reception.

Now, let's look at the materials you'll need, the steps you need to follow, and some practical tips that'll help you along the way.

Materials Needed

- **Wire:** The primary material for the antenna elements. For most field applications, stranded wire is preferred for its flexibility and durability, but solid wire can also be used if necessary.

- **Insulators:** Can be anything from commercial antenna insulators to homemade alternatives like plastic tubing, sections of PVC pipe, or even dry wooden blocks.

- **Measuring Tape:** Essential for accurately measuring the wire lengths.

- **Wire Cutters and Strippers:** For cutting the wire to length and stripping insulation where connections will be made.

- **Coaxial Cable:** To feed the antenna from the radio. The length may vary depending on the specific setup requirements.

Step-by-Step Construction

- **Determine the Antenna Design:** Decide on the type of antenna you need based on your communication requirements – whether it's a simple dipole for broad coverage or a directional antenna like a Yagi for focused communication.

- **Calculate Element Lengths:** Use the formula $\lambda = cf$ to calculate the wavelength (λ) for your desired frequency (f), where c is the speed of light (approximately 3×108 3×108 m/s). For a dipole, each element should be $\lambda 44\lambda$ in length.

- **Cut and Prepare the Wire:** Based on your calculations, cut the wire to the necessary lengths for your antenna elements. Remember to account for a small extra length for making connections.

- **Attach Insulators:** Attach insulators to the ends of each antenna element. This could be as simple as tying the wire through holes drilled in small plastic or wooden blocks.

- **Assemble the Antenna:** Depending on the antenna type, assemble the elements into the required configuration. For a dipole, this might involve securing the center point where the two halves meet and where the coaxial cable will connect.

- **Connect the Coaxial Cable:** Strip the end of the coaxial cable and attach it to the antenna. The center conductor attaches to one side of the antenna element, and the shield to the other. Ensure a secure and weatherproof connection.

- **Test and Adjust:** Once assembled, test the antenna with your radio. Minor adjustments to the element length may be needed to fine-tune performance.

Practical Tips

- **Use What's Available:** In field conditions, adaptability is key. Don't hesitate to use unconventional materials as insulators or supports if they're non-conductive and available.

- **Test with Tools:** If possible, use an SWR meter to check the antenna's performance and make adjustment accordingly.

- **Keep Safety in Mind:** Always erect antennas away from power lines and in locations where they won't pose a hazard to people or wildlife.

Antenna Orientation

The next thing you need to focus on is orientation. For this, you'll need to decide between whether the antenna should be vertical and horizontal polarization. We'll cover both of these in detail in a second, but for now just know that they can impact communication, signal transmission, and security. We'll start with vertical polarization first and cover its practical implications and benefits and we'll do the same for the other one.

Vertical Polarization

Vertical polarization means the antenna radiates energy in an up-and-down pattern. This is the most common orientation used for line-of-sight (LOS) radios, including handheld and mobile units like the Baofeng radios discussed in this manual. It's quite popular because of its efficiency in transmitting signals across the earth, which makes it ideal for communication. The practical implications and benefits of vertical propagation include:

Practical Implementation

- **Handheld Radios:** The antennas are naturally oriented vertically when held upright, aligning with the standard use case.

- **Mobile Units:** Vehicles' antennas should be mounted vertically to maintain consistent polarization with other radios in the communication network.

Benefits

- **Compatibility:** Most LOS communications are designed with vertical polarization in mind, ensuring compatibility across devices.
- **Terrain Adaptability:** Effective in a wide range of environments, particularly where the ground can reflect radio waves upward.
- **Simplicity:** Natural for handheld use without requiring specific adjustments for orientation.

Horizontal Polarization

Horizontal polarization, on the other hand, means the antenna radiates energy in a side-to-side pattern. This type of polarization is less commonly used in standard LOS communication. But that doesn't mean it's not worth considering. It has some distinct uses and benefits that include:

Practical Implementation

- **Strategic Placement:** When security or stealth is a priority, antennas may be oriented horizontally to reduce the signal's detectability from certain angles.
- **Environment Matching:** In environments with significant horizontal structures, such as urban settings with tall buildings, horizontal polarization can sometimes provide better signal penetration.

Benefits

- **Security:** Horizontal polarization can reduce the likelihood of signal interception by vertically polarized receivers, offering an additional layer of communications security.
- **Reduced Interference:** In certain environments, horizontal polarization can help minimize interference from other vertically polarized signals.

Choosing the Right Orientation

Alright, so now that you're well aware of what both these polarizations are, you need to choose one. To do this, consider the following:

- **Operational Security:** Horizontal polarization might be preferred in covert operations where minimizing signal detection is critical.
- **General Use and Compatibility:** Vertical polarization is usually the best choice for general communication needs due to its compatibility and effectiveness in a broad range of environments.
- **Environment-Specific Needs:** Consider the operating environment's characteristics, such as urban vs. rural, flat vs. varied terrain, and the presence of natural or man-made structures that could impact signal propagation.

Field Expedient Antenna Construction

Remember when we talked about dipole antennas? Great! Because now, it's DIY time. In this section, we'll look at how we can create three different antennas. These include:

- Dipole antenna.

- Inverted V Dipole antenna.

- Jungle antenna.

Let's begin!

DIY Dipole Antenna: Enhancing Baofeng Radio Performance

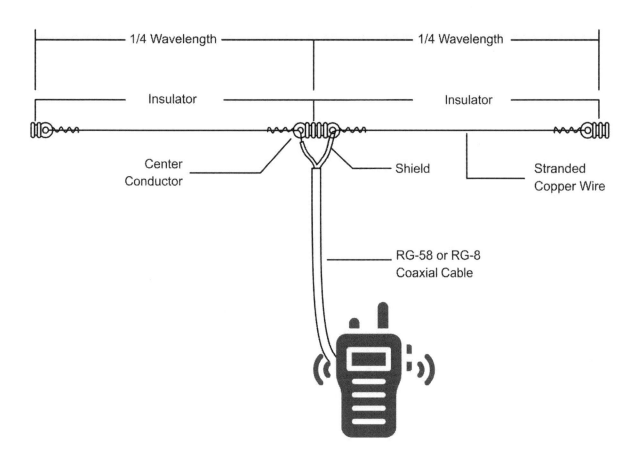

Creating a dipole antenna for ham radio is an exciting DIY project. Believe it or not, it's both cost-effective and there are lots of customization options. Some of key benefits of this type of an antenna include:

- Customization: Tailor your antenna to the exact frequency you desire.
- Cost-Effectiveness: With a single roll of speaker wire, you can create multiple antennas.
- Choice of Balun: Select or build a balun that best suits your specific needs.
- Practical Experience: Gain valuable hands-on experience in antenna construction.

Now to create such an antenna, you will need to have the following materials: :

- Roll of speaker wire
- Measuring and cutting tools
- Balun (either commercial or homemade)
- Insulators

Once you have all these items, just follow the steps mentioned below:

Calculating Dipole Length

- Use the formula: 468 / Frequency (MHz) = Length of each side of the dipole (in feet).
- For a 20M dipole at 14.250 MHz, each side should be approximately 16.5 feet long. For a 40M dipole, each side should measure around 32.5 feet.

Measuring and Cutting the Wire

- Measure the required length for your dipole. For a 20M dipole, you'll need 16.5 feet of speaker wire. Since speaker wire is dual conductor, splitting it will give you two equal lengths for each side of the dipole.

Preparation for Tuning

- Leave a little extra wire for tuning adjustments. It's easier to trim excess wire than to add more if you cut it too short.

Connecting the Wire to a Balun

- Strip the ends of the wire if necessary and attach it to the terminals of the balun. The balun transforms balanced antenna signals to unbalanced feedline signals, which can then be connected to your radio.

Final Preparations Before Deployment

- Attach dog bone insulators at the ends of each dipole arm for insulation. This is especially useful when attaching the antenna to supports like trees or poles.
- For a 40M dipole, you might opt for a homemade balun, utilizing a 3D printed casing and an SO239 connector for easy connection to the antenna wires.

Inverted V Dipole Antenna: Quick-Construction Directional Antenna Detailed Explanation

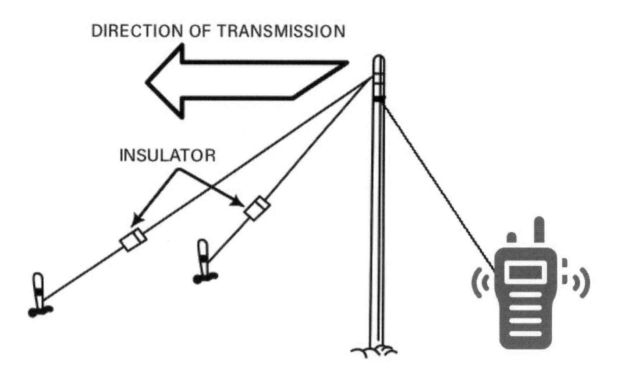

Next up on our list is the Inverted V antenna. It's a practical antenna that you can make pretty quickly by using just a few things. This antenna will help you direct communication efforts in a specific direction. It's a simple yet effective way of directing radio communications to a particular location or receiver and can be used for field operations.

Since it's focused on a targeted direction, it can give you better range and signal strength. When you're using the Inverted V antenna, remember the open side of the V should face towards the target area or direction of communication. Some key benefits of this type of antenna, include:

● Simplicity: Easy to construct with readily available materials.
● Directionality: Enhances communication in a specific direction, improving signal strength and range.
● Flexibility: Can be quickly reoriented or adjusted to meet changing operational requirements.
● Portability: Lightweight and does not require permanent installations, making it ideal for mobile units or temporary setups.

Now, let's look at the materials that are required and the steps you need to follow to assemble the antenna.

Construction Details: Materials Required

- **Two pieces of wire:** Each piece should be half a wavelength long at the desired frequency of operation. The wavelength (λλ) can be calculated using the formula λ=cf, where c is the speed of light (3×1083×108 m/s) and f is the frequency in hertz.

- **Insulators:** Needed for securing the ends of the wires and preventing unwanted electrical contact with the ground or other objects.

- **Support Structure:** Something to elevate the center of the antenna, such as a tree, pole, or any available high point.

- **Coaxial Cable:** Used to connect the antenna to the radio, transmitting the received or broadcast signals.

Assembly Steps

- **Central Support:** Begin by securing the midpoint of one of the wires to your high point using an insulator. This point acts as the apex of the Vee.

- **Forming the Vee:** Stretch each end of the wire downwards and apart, creating a V shape. The angle between the two halves of the wire should ideally be around 90 degrees (45 degrees to the ground from each wire), but there is flexibility based on environmental constraints.

- **Ground Connection:** Use insulators at the end of each wire to secure them to stakes in the ground or other non-conductive anchor points.

- **Connection to Radio:** The coaxial cable from the radio connects to the antenna at the apex. The center conductor of the coax attaches to one wire, and the shield (outer conductor) attaches to the other. This configuration ensures that the antenna is properly fed with the radio signal.

Jungle Antenna: Omni-directional RF Range Enhancement Detailed Explanation

Alright, so now we're close to the end of this chapter. The last type of DIY antenna we're going to cover is the Jungle antenna. If you're someone who's a bit tech savvy, you can call it the Quarter Wave Ground Plane. This antenna is specifically designed to enhance radio frequency (RF) range in an omni-directional manner.

What this means is that it's exceptionally suited for use in an environment where wide-range communication is needed. It'll come in handy when you're at a base camp, near relay points, or at emergency communication setups. So, with that in mind, let's look at the materials and tools you need and how you can assemble and enhance this antenna.

Required Materials:

- **Split post BNC adapter** (also known as "Cobra Head")
- **Four ring terminals** that fit the BNC adapter posts
- **Wire** about 10 feet long (e.g., 20 AWG primary wire, 14-gauge THHN, or lamp cord)
- **Four insulators** (such as zip ties, shoelaces, electric fence insulators, or cut PVC pipe)
- **550 cords** for hanging the antenna over a tree branch
- **Three straight sticks** about 2.5 feet long and as thick as your finger

Required Tools:

- Tape measure
- Wire cutters
- Wire strippers
- Crimping tool

Assembly Procedure:

- **Determine the Desired Frequency:** For example, MURS 3 (151.940 MHz).
- **Calculate Each Antenna Leg Length:** Use the formula Length in feet=234 frequency in MHz. Length in feet=frequency in MHz 234.
  - For MURS 3, this results in 1.54 feet.
- **Convert to Inches and Add Excess:** Convert the length to inches and add 5.5 inches for crimping and the insulator loop.
  - Result: 18.5 inches+5.5 inches=24 inches 18.5 inches+5.5 inches=24 inches.
- **Cut Four Lengths of Wire:** Measure and cut the wires to the obtained length.
- **Prepare Wires for Assembly:** Mark each wire 5 inches from one end for the excess, strip half an inch from the other end, and crimp on the ring terminals.
- **Assemble the Antenna Legs to the Cobra Head:** Attach one leg to the red side (radiator/hot) and the other three to the black side (ground/cold).
- **Prepare the Antenna's Support:** Arrange the three sticks on the ground in a triangle shape to support the ground legs angled at 45 degrees.
- **Hang the Antenna:** Use the 550 cord to hoist the antenna over a branch, connect to the radio via a coaxial cable, and ideally position the antenna away from live trunks and branches to reduce interference.

Enhancements:

- **To Increase Gain:** Adjust the lengths of the three ground legs to be 12% longer.
- **To Minimize Losses:** Use higher quality cables for longer connections and adjust the leg lengths to optimize the SWR (Standing Wave Ratio).

SWR (Standing Wave Ratio) for Antenna Efficiency

If you're thinking that you are now a DIY antenna expert, you're almost right. Just a couple of things left to cover. Now, you already know how to make different types of antennas, but just knowing how to make them isn't going to do you any good if you can't improve their efficiency, right?

That's where the Standing Wave Ratio (SWR) comes in!

The SWR is a crucial parameter that indicates how efficiently an antenna is able to transmit radio frequencies. It compares the amplitude of a partial standing wave at its maximum to the amplitude at its minimum. What this basically means is that it represents the ratio of the power sent to the antenna to the power reflected back from the antenna.

Knowing how to monitor this ratio can help you prevent your equipment from getting damaged and allows you to improve your radio communications. Let's look at both these in a bit more detail.

- Preventing Radio Damage: A high SWR indicates that a significant amount of power is being reflected back into the transmitter rather than being radiated by the antenna. This reflected power can cause overheating and damage the radio's final amplifier stage.
- Ensuring Efficient Communication: The closer the SWR is to 1:1, the more efficiently the antenna system operates. An efficient antenna system means better transmission and reception of signals, which is crucial for clear and effective communication.

Now you also need to know that an SWR of 1:1 is considered ideal. It indicates that all the transmitted power is radiated by the antenna, with no reflection. However, achieving a perfect 1:1 SWR is often challenging.

So, an SWR below 2:1 is considered safe and indicates reasonable efficiency. What this means is that most of the power is being transmitted, and the risk of equipment damage is low.

To measure SWR, you'll need an SWR meter, which is connected between the transmitter (radio) and the antenna. Here are the basic steps to measure SWR:

1. **Connect the SWR Meter:** Connect the radio to the SWR meter's input labeled as "TX" or "Radio" and the antenna to the output labeled as "ANT."
2. **Set the Radio to Transmit Mode:** Choose a frequency in the middle of the band you're using for your communication.
3. **Transmit and Read the Meter:** Activate the transmit mode on your radio (often by pressing the push-to-talk button) and observe the SWR reading on the meter.
4. **Adjust the Antenna as Necessary:** If the SWR reading is above 2:1, adjustments to the antenna length or position may be necessary to improve the match and lower the SWR.

Once you've measured the SWR, here are some of the ways you can optimize it:

- **Antenna Tuning:** Many antennas come with adjustable elements for tuning purposes. Adjust these elements as per the manufacturer's instructions to improve SWR.
- **Antenna Placement:** Keep the antenna away from large metal objects and adjust its position to find the spot where SWR is the lowest.
- **Use Quality Coaxial Cable:** A high-quality coaxial cable with the correct impedance for your system (usually 50 ohms) can reduce SWR.
- **Regular Checks:** Regularly check your SWR, especially if you have changed your antenna's location, the surrounding environment has changed significantly, or after any modifications to your antenna system.

Portable Antenna Solutions

If you're someone who loves outdoor adventures but still wants to stay connected from remote locations, then portable antenna solutions are something worth considering. The purpose of portable antenna solutions is that you have access to reliable communication in remote or challenging environments.

Be it hiking, camping, or a wilderness expedition, with a portable antenna on your side you'll stay connected with your team, receive critical weather updates, and even call for help in emergencies.

However, when choosing a portable antenna for your Baofeng radio, you need to think about different factors such as size, weight, and antenna type. Now, let's look at some popular antenna options and set up protocols before we wind up this chapter.

Popular Portable Antenna Options

- **SMA-Female Dual Band Antenna**: This compact antenna is compatible with most Baofeng radios and offers reliable performance across VHF and UHF bands.

- **Telescopic Whip Antenna**: With its collapsible design, the telescopic whip antenna provides excellent portability and can be extended for enhanced signal reception.

- **Wire Antenna with BNC Connector**: A simple wire antenna with a BNC connector offers flexibility and can be easily deployed in outdoor settings by hanging it from trees or poles.

Setting Up Your Portable Antenna

- **Attach the Antenna**: Connect the portable antenna to your Baofeng radio by screwing it onto the antenna connector. Ensure a secure connection to optimize signal transmission.

- **Extend or Deploy the Antenna**: Depending on the antenna type, extend or deploy it to its full length for optimal performance. Position the antenna vertically for maximum signal strength.

- **Select the Operating Frequency**: Use the radio's frequency settings to tune into the desired frequency band for communication or monitoring.

CHAPTER 12:
Advanced Programming Techniques

Remember when we learned about how you can program your radio? Well, in this chapter, we're going to take a trip back to that part and cover some advanced programming techniques. We'll look at how you can customize your Baofeng radio to your unique needs, channel management and alpha tags. So, without further ado, let's begin!

Tailoring Baofeng for Specialized Needs

Tailoring your Baofeng radio to meet your customization needs is actually quite

simple. All it takes is a few adjustments here and there. These adjustments might appear to be small, but they go a long way in optimizing the functionality. You can count on them regardless of whether you're a hobbyist, adventure lover or an emergency responder. Here are some customization options and implementation tips that'll help you out.

Customization Options

1. **Channel Programming**: Organize and prioritize channels based on frequency, location, or function. By programming channels according to your specific needs, you can quickly access essential frequencies without scrolling through irrelevant channels. To program a channel, press the **MENU** button, navigate to the channel programming menu, and enter the desired frequency using the numeric keypad. Press the **MENU** button again to save the settings.

2. **VOX Sensitivity Adjustment**: Voice-activated transmission (VOX) allows hands-free operation, making it ideal for certain scenarios such as driving or multitasking. Adjusting the VOX sensitivity ensures accurate voice detection and transmission. To adjust VOX sensitivity, press the **MENU** button, navigate to the VOX settings

menu, and select the desired sensitivity level. Higher sensitivity levels detect softer sounds, while lower levels require louder input.

3. **Scanning Preferences**: Customize scanning parameters to focus on specific frequency ranges or skip unwanted channels. Tailoring scanning settings streamlines the monitoring process and ensures efficient frequency surveillance. To customize scanning preferences, press the **MENU** button, navigate to the scanning settings menu, and adjust parameters such as scan mode, scan resume options, and priority channel settings.

Implementation Tips

4. **Identify Your Needs**: Assess your communication requirements and identify areas where customization can enhance performance and efficiency.

5. **Experiment and Test**: Don't hesitate to experiment with different settings and configurations to find the optimal setup for your needs. Testing your customized settings in various scenarios ensures reliability and functionality.

6. **Document Your Setup**: Keep a record of your customized settings and configurations for future reference. Documenting your setup streamlines the customization process and allows for easy replication or adjustment as needed.

Custom Channels

You can use your Baofeng radio's settings to craft and manage channels based on what you like and don't like. This can help you organize and prioritize the frequencies and allows you to streamline the communication process. Here's what you need to do to craft and manage custom channels:

Crafting Custom Channels

1. **Accessing the Menu**: To begin crafting custom channels, press the **MENU** button on your Baofeng radio.

2. **Navigating to Channel Programming**: Scroll through the menu options using the arrow keys until you find the channel programming or frequency settings.

3. **Entering Frequency**: Once in the channel programming menu, use the numeric keypad to enter the desired frequency for your custom channel. Press the **MENU** button again to save the frequency.

4. **Assigning Name (Optional)**: Some Baofeng models allow you to assign names to custom channels for easier identification. If available, navigate to the channel naming option and enter a name using the alphanumeric keypad.

5. **Saving the Channel**: After entering the frequency and optional name, save the custom channel settings by pressing the **MENU** button or following on-screen prompts.

Managing Custom Channels

6. **Editing Channels**: To edit or modify existing custom channels, access the channel programming menu and navigate to the desired channel. From there, you can adjust the frequency, name, or other settings as needed.

7. **Deleting Channels**: If you no longer need a custom channel, you can delete it from your radio's memory. Navigate to the channel programming menu, select the channel you wish to delete, and follow the on-screen prompts to remove it from the list.

8. **Organizing Channels**: Organize your custom channels by grouping them based on frequency range, location, or function. This makes it easier to navigate through channels and access the ones you need quickly.

Channel Identification with Alpha Tags

Ever feel tired of scrolling through different frequencies with hopes of finding the right one? Been there, felt that! But you don't have to anymore if you start using Alpha tags. These tags are a simple and effective way for you to assign names to your custom channels which makes it easier for you to identify them later on. To do this, just follow the steps mentioned below:

1. **Accessing Menu**: Press the **MENU** button on your Baofeng radio to enter the menu mode.

2. **Navigating to Channel Settings**: Scroll through the menu options using the arrow keys until you find the channel settings or channel mode.

3. **Selecting Channel**: Choose the channel you want to assign an alpha tag to by using the arrow keys or numeric keypad.

4. **Entering Alpha Tag Mode**: Once you have selected the desired channel, enter the alpha tag mode by pressing the **MENU** button or a designated function key.

5. **Entering Alpha Tag**: Use the keypad to input the custom name or label for the channel. Navigate through the alphanumeric characters using the arrow keys and confirm each character by pressing the corresponding numeric keypad button.

6. **Saving Alpha Tag**: After entering the desired alpha tag, save the settings by pressing the **MENU** button or following any on-screen prompts.

CHAPTER 13:
Expanding Baofeng's Connectivity

In this chapter, we'll talk about the Baofeng radio's connectivity and how you can expand it! Yeah, that's right you can expand its connectivity, too. You can connect it to external speakers, GPS devices, mobile phones.

You can even use cable and Bluetooth for increased connectivity options. Eager to know who you can do all of that. Let's begin.

Integrating Baofeng with External Speakers

First up, let's talk about external speakers because, well, who doesn't like better clarity and more volume, right? Remember that remote island example we went over earlier in the book? Great!

So, you're on that island, have a bonfire set up and are planning some barbeque. What could possibly amp up the ambiance? Music, you guessed it! To get that music going you can connect your Baofeng radio to external speakers.

But this integration isn't just good for music alone. It can help you with other outdoor activities and emergency situations as well. To connect your Baofeng radio to external speakers, just follow the steps below:

1. **Locate Audio Output Jack**: Identify the audio output jack on your Baofeng radio. It is typically located on the side or back of the device and labeled as "AUDIO OUT" or "EARPHONE."

2. **Select Compatible Speaker**: Choose an external speaker compatible with your Baofeng radio. Ensure that the speaker has a matching audio input jack and is suitable for the intended environment (e.g., portable, waterproof, etc.).

3. **Insert Audio Cable**: Connect one end of the audio cable into the audio output jack of your Baofeng radio.

4. **Connect to External Speaker**: Plug the other end of the audio cable into the audio input jack of the external speaker.

5. **Power On**: Turn on your Baofeng radio and the external speaker. Make any necessary adjustments to the volume settings on both devices.

Connection to GPS Devices

Alright, so let's say that things don't turn out so good on the island. You got stranded, some of your friends or maybe even you yourself lost your way, or something completely different happened. The point is, now you need to find your way to a specific destination.

What are you going to do in such circumstances? You guessed it. The answer is: you're going to connect your Baofeng radio to a GPS device. This will allow you to transmit and receive location data which you can use to map your way to wherever it is you need to go. To connect the Baofeng radio to a location device, here's what you have to do:

1. **Choose Compatible GPS Device**: Select a GPS device that is compatible with your Baofeng radio. Ensure that the GPS device supports data transmission via audio cable or Bluetooth connection.

2. **Connect Audio Cable**: If you are using an audio cable connection, locate the audio input/output jacks on both the Baofeng radio and the GPS device. Connect one end of the audio cable to the audio output jack of the GPS device and the other end to the audio input jack of the Baofeng radio.

3. **Enable Data Transmission**: On your Baofeng radio, navigate to the settings menu and enable data transmission mode. This mode allows the radio to send and receive location data through the connected GPS device.

4. **Configure GPS Settings**: Configure the GPS settings on your Baofeng radio to specify the data format and transmission frequency. Ensure that the settings match those of the connected GPS device for seamless communication.

5. **Test Communication**: Perform a test transmission to verify that the Baofeng radio is successfully sending and receiving location data from the GPS device. Confirm that the GPS coordinates are accurately displayed on the radio screen.

Bluetooth Connectivity

Next up on our list is Bluetooth! Some might say that it's an outdated technology and connectivity option. But, let me tell you that knowing how to connect your Baofeng radio to a Bluetooth device is worth it.

Whether you're using a Bluetooth headset for hands-free operation, streaming audio through a Bluetooth speaker, or connecting to a smartphone for data transfer, Bluetooth connectivity enhances your radio experience. To pair the Baofeng radio to different devices that support Bluetooth, you need to:

1. **Enable Bluetooth Mode**: Turn on the Bluetooth mode on your Baofeng radio by navigating to the settings menu and selecting the Bluetooth option. This activates the radio's Bluetooth functionality and allows it to search for nearby devices.

2. **Put Bluetooth Device in Pairing Mode**: If you're pairing your Baofeng radio with a Bluetooth headset, speaker, or smartphone, ensure that the device is in pairing mode. Refer to the device's user manual for instructions on how to activate pairing mode.

3. **Search for Devices**: On your Baofeng radio, initiate a search for nearby Bluetooth devices by selecting the "Search" or "Scan" option in the Bluetooth menu.

4. **Select and Pair Device**: Once your Baofeng radio detects the Bluetooth device, select it from the list of available devices and initiate the pairing process. Follow the on-screen prompts to complete the pairing procedure.

5. **Confirm Connection**: After successfully pairing your Baofeng radio with the Bluetooth device, confirm the connection status on both devices. You should see a notification indicating that the devices are connected.

Data Modes and Connections

When it comes to radio communication, most people just think that it's all analog. But you can use data modes or connections and use your Baofeng radio for digital communication and data transfer. This allows you to transmit GPS coordinates, text messages, images, and other data formats.

All of this can be quite handy for both casual conversations and emergency situations. To make such a connection, here's what you need to do:

1. **Select Data Mode**: Begin by selecting the desired data mode on your Baofeng radio. Common data modes include APRS (Automatic Packet Reporting System), SSTV (Slow Scan Television), and PSK31 (Phase Shift Keying 31). Navigate to the radio's menu and choose the appropriate data mode option.

2. **Connect Data Cable**: If connecting your Baofeng radio to a computer or external device for data transmission, ensure that the appropriate data cable is connected to the radio's data port. Use a USB cable or audio cable depending on the type of connection required.

3. **Configure Settings**: Adjust the settings on your Baofeng radio to match the requirements of the selected data mode. This may include setting the modulation type, baud rate, frequency deviation, and other parameters specific to the chosen data mode.

4. **Initiate Transmission**: Once the settings are configured, initiate transmission by pressing the push-to-talk (PTT) button on your Baofeng radio. The radio will transmit data packets containing the information you wish to send, whether it's a text message, image, or telemetry data.

5. **Monitor Reception**: Monitor the reception of data packets on your Baofeng radio or connected device. Depending on the data mode used, you may need to decode received packets using compatible software or hardware decoders.

Integration with Tablets or Mobile Devices

An interesting thing about the Baofeng radio is that you can connect to tablets and mobile devices, too. This gives you access to a wide variety of digital communication options and allows you to ensure secure data transfer options. But, to do this, you'll need to use the andFLmsg app. We'll talk more about that in a second. But, first let's look at some of the key reasons for connecting your Baofeng radio to such devices. These reasons include:

- Enhanced Security: Digital modes, combined with encryption, make it significantly harder for unauthorized parties to intercept and understand communications.
- Improved Accuracy: Sending text or data ensures precise communication, reducing errors associated with voice messages, especially in noisy or stressful environments.
- Increased Efficiency: Digital data transmission can be quicker and more reliable over distances, especially for sending complex or structured information like coordinates or pre-defined messages.

Alright so the andFLmsg is a software application that is made for digital communication through amateur radio frequencies. It allows you to send and receive various types of data, including text messages, emails, and forms over radio waves. The application can facilitate data transmission in multiple formats by encoding the data in audio signals.

You need to know that these formats can be more efficient and secure compared to standard voice communication. Such an integration is particularly useful in scenarios where internet or cellular networks are not available. Common examples of such scenarios may include emergency situations, remote locations, or for coordinating activities among amateur radio enthusiasts.

So, with that in mind, let's look at the equipment you need, the setup process, and a few other useful things.

Preparation: Equipment Needed

- **Baofeng Radio:** Any model that supports voice operation over the required frequencies.
- **Tablet or Mobile Device:** With andFLmsg app installed.
- **APRS Cable:** Specifically designed to connect your Baofeng radio with your tablet or mobile device.

Step-by-Step Setup

- **Install andFLmsg App:** First, ensure that the andFLmsg application is installed on your tablet or mobile device. This app is available for free and supports a variety of digital modes.
- **Connect APRS Cable:** Connect one end of the APRS cable to the headphone/microphone jack of your Baofeng radio. The other end of the cable should be plugged into the audio jack of your tablet or mobile device.
- **Configure Radio Settings:**
  - Turn on your Baofeng radio and ensure it's set to **VFO Mode**.
  - Adjust the **Volume** to about two-thirds for optimal audio transfer.
  - **Enable VOX** (Voice Operated Exchange) by accessing the menu (MENU #4) and setting it to the lowest trigger level. This allows the tablet to key the transmitter automatically when data is being sent.

Configure andFLmsg:

- Open the andFLmsg app on your device.
- Choose the appropriate **Digital Mode** for your operation. MT63 2000L is recommended for its balance of speed and reliability.
- **Compose a Message** or select a **Form** to fill out if sending structured information.

Test the Connection:

- Before proceeding with actual operations, conduct a test transmission to ensure everything is configured correctly. Send a simple message from the app and verify it is received on another radio set to the correct frequency and mode.

Key Points for Operation

- **Minimal Transmission Time:** Keep digital transmissions as brief as possible to reduce detection risk.
- **Frequency Selection:** Use frequencies that are less likely to be monitored by adversaries and ensure they are compatible with your digital mode.
- **Operational Security:** Encrypt sensitive information before sending it digitally. Although digital modes offer an additional layer of security, encryption adds an essential barrier against interception.

Practical Use Cases

- **Sending Encrypted Texts:** Quickly transmit encrypted messages containing orders, intelligence, or coordination instructions.
- **Sharing GPS Coordinates:** Send exact locations for rendezvous points, caches, or targets without speaking to them out loud.
- **Emergency Communications:** Transmit pre-formatted messages for distress signals, medical emergencies, or extraction requests.

Part II - Baofeng in Emergency and Survival Scenarios

C H A P T E R   1 4 :
Guerrilla Communications

Before we begin part two of this book let's just take a quick look back and see what we've learnt. In part one, we basically covered things like antennas, signals, tone, frequencies, and pretty much all the other features and functionalities of the Baofeng radio.

So now that you're well versed with all that, we're going to talk about how to master communication in emergency situations using your Baofeng radio. You see, the thing with emergency situations is that you can go from a controlled to a chaotic environment quite quickly.

What can you do to communicate effectively in such scenarios? That's exactly what we're going to cover in this chapter.

Strategic Roles of Communications

Before we get into how you can use your Baofeng radio for guerilla communications, first you need to know what the different roles of communication are and how they can vary from one another.

To start off, you need to know that the roles of communication are categorized into three different types. This includes sustainment, tactical, and strategic communication. Let's uncover each of them one at a time.

Sustainment Communications

I know this might seem like a difficult thing to understand based on how the name sounds, but it's not. Sustainment communication basically means regular everyday communication that you'd have but in cases where traditional communication methods aren't working. You know, like times when cellphone networks and the internet are down.

The end goal here is to have a reliable way of communicating things to such vital information, updates, and relie requirements within a group. This can be quite handy in times of emergencies or isolation. One key thing to remembe here is that with sustainment communication, the priority here is not to make the communications secure, but to make them clear and effective.

Tactical Communications

Now, when it comes to tactical communication, you need to know that these types of communication efforts are mainly used in military-related operations. The key focus here is on ensuring that you can transmit information regarding movements and actions since it can help keep everyone safe.

With tactical communications, a higher level of security is needed. Due to this, such communication efforts often use codes, encryption, and low-power transmissions. Why is that the case? Well, you see such communication protocols help lower the risk of detection or interception. Oh, and one more thing: to make tactical communication effective you need to know the ins and outs of the operational environment and the equipment.

Clandestine/Strategic Communications

For this one what you need to know is that strategic communications are often used in environments where security concerns are higher. This could be during an undercover operation, maybe a coup, or when operations are being conducted in enemy territory. Strategic communication is used when messages or information needs to be transmitted over a long distance. .

As far as equipment and protocols are concerned this type of communication often uses things like encryption and directional antennas. This ensures that unauthorized individuals cannot understand or use the data being transmitted and that it cannot be intercepted.

Communications Security and Operational Best Practices

Alright, so now that you're aware of the basic types of communication, let's look at some security and operational best practices. The first thing you need to know is that in tactical and strategic communications security protocols are critical. But, with sustained communication, the focus is on clarity and effectiveness.

Now the next thing you must be aware of is that one of the most critical aspects of communication security, also known as COMSEC, is managing sensitive frequencies. The great thing here is that you can use your Baofeng radio to manage these frequencies by programming and storing them in its memory. Although this feature is effective, it does come with some risk.

Let's say that you had some frequencies stored on your Baofeng radio and they were used to transmit top secret information. Now, this wouldn't be a risk as long as you were in possession of the radio. But, if it were to be stolen unauthorized individuals could easily access the information. In tactical or strategic communications this could lead to details of strategic locations, operations, and future plans being compromised.

So, if you plan to use the Baofeng radio for tactical or strategic purposes and have to store frequencies on them, keep all of this in mind. Some of the things that you can do to mitigate such risks and improve security include:

1. **Use Frequency Mode for Operations:** Instead of storing frequencies in the radio's memory, operators should use the radio in Frequency Mode. This requires manually entering the frequency for each operation, which, while potentially less convenient, significantly reduces the risk of sensitive information being easily accessible in the event of capture.

2. **Regularly Change Frequencies:** Even without storing them in memory, it's critical to regularly change operational frequencies to avoid pattern recognition by adversaries equipped with direction-finding technology or signals intelligence capabilities.

3. **Minimalist Programming:** For non-sensitive or public service operations (e.g., coordinating disaster relief efforts), limit programming to non-operational frequencies or those widely known and used for public safety and service.

4. **Operational Discipline:** Train all operators in strict COMSEC practices, including the physical security of radios, the importance of not discussing sensitive information over unsecured channels, and the procedures for the destruction or disposal of equipment that may be compromised.

5. **Factory Resets as a Precaution:** In the event of potential compromise, utilizing the factory reset function on the Baofeng radio can erase stored channels and settings, though this action should be considered a last resort due to its overt nature.

Ensuring Tactical Integrity through Keypad Lock Function

Another thing you can do to ensure communication security during tactical and strategic operations is to use the keypad lock feature on your Baofeng radio. This is one of the most important features when it comes to ensuring communication security. But since it's a minor feature it can be overlooked by many.

The thing you need to know about the Baofeng radio's keypad lock feature is that it's designed to keep certain functions from going off simply because a button got pressed accidentally. This neat little feature is very important in open-field conditions where you're constantly moving around and always have the radio with you. Think about it for a second. What would happen if a button got pressed accidentally?

Frequency change or mode change? FM radio going off? Flashlight turning on? Or all the above? The truth is you never really know what could happen. This could even damage communication links and that is something which can be disastrous, right?

So, to keep any of this from happening, here's what you can do:

1. **Activate the Lock:** Simply press and hold the # key. The radio will confirm the activation of the lock with a voice prompt, stating "Lock," if voice prompts are enabled. Additionally, a lock icon will appear on the display indicating that the keypad lock is engaged.

2. **Understanding What's Locked:** With the keypad lock active, all front panel buttons except for the Push-To Talk (PTT) button are disabled. This means that volume adjustments and emergency alerts (if accidentally triggered by side buttons) are still possible, emphasizing the need for careful handling even when the keypad is locked.

3. **Deactivating the Lock:** To unlock the keypad, press and hold the # key again until the radio announces "Unlock," and the lock icon disappears from the display. The keypad will then return to its normal operational state.

4. **Operational Discipline:** Operators should develop the habit of engaging the keypad lock immediately after setting their frequency and other necessary adjustments. This discipline should be ingrained through training and practice, ensuring it becomes second nature, akin to safety checks on a weapon.

5. **Volume Control and Alerts:** It's crucial to remember that the keypad lock does not affect the volume control knob or prevent the activation of the flashlight or alarm (through side button long presses). Operators should remain cognizant of these aspects to avoid inadvertent noise or light discipline breaches.

Communication Principles for Unconventional Warfare

Now, what we are about to cover might not be directly relevant to you, but it's still worth learning. You know that Baofeng radio can be used during warfare and in such environment communication efforts need to be disciplined and efficient. There are a few key principles that can help ensure such discipline. Let's go over each of them and look at a few practical examples.

Principle 1: Brevity is Key

Keep Transmissions Short: Each message should be as brief as possible. This not only reduces the window of opportunity for an adversary to detect and intercept the communication but also aids in the clarity of the message. For example instead of saying, "Team Bravo, please advance towards the eastern ridge line and report any sightings of enemy patrols," use condensed phrasing such as, "Bravo, move east ridge, report enemy."

Principle 2: Clarity Cannot be Compromised

Ensure Clear, Unambiguous Messages: Every transmitted word should be chosen for its clarity and unambiguity. Use established code words or phrases when possible, and avoid slang or local dialects that could confuse recipients. For instance, instead of using colloquial terms for locations or actions, stick to pre-agreed codes such as "Rally Point Alpha" or "Execute Plan Delta."

Principle 3: Assume You're Being Listened To

Operate as if Every Transmission is Monitored: Always assume that the enemy is listening. This mindset should influence not only the content of what is communicated but also how and when you transmit. For example, instead of specifying exact times or locations, use relative terms or pre-established codes understood by your team but meaningless to anyone else. Instead of saying, "Meet at the abandoned factory at 0400 hours," say, "Rendezvous at Location Zulu at minus two hours."

Practical Examples and Implementations:

- Example 1: Coordinating an Ambush
  - Poor Practice: *"All units, prepare to engage the enemy convoy when it turns onto Highway 22. Wait for my command."*
  - Best Practice: *"All units, standby. Execute on mark."*
- Example 2: Reporting an Observation
  - Poor Practice: *"I've spotted about a dozen enemy troops moving through the forest, heading towards our main camp. They look heavily armed."*
  - Best Practice: *"Eagle, see twelve, moving towards Home. Alert."*
- Example 3: Requesting Support
  - Poor Practice: *"Can someone bring more ammunition to the southern checkpoint? We're running low and might not hold out if they attack again."*
  - Best Practice: *"Supply, need ammo South Check. Urgent."*

Prowords and Phonetic Alphabet

For this next part, we're going to cover Prowords and the NATO phonetic alphabet. These are tools that help you make the most of your communication efforts as they can minimize misunderstanding and make sure that the message is conveyed across accurately. Let's look at what both of these things are.

Prowords are special words that have a specific meaning and are used to prevent any confusion. Some common examples of these words may include:

- OVER: Indicates that the sender has finished speaking and is waiting for a reply from the receiving station.
- OUT: Signifies that the sender has finished the transmission and does not expect a reply.

- ROGER: Acknowledges that a message has been received and understood.

You've probably heard these words being used in a movie or a TV season, but here are a few practical examples that help you get a clear idea.

Practical Example 1: Coordination of Movement

- Incorrect Use: "I have finished my message. Do you understand and have anything to add?"
- Correct Use: "Do you copy my last, OVER?"

Practical Example 2: Confirming Orders

- Incorrect Use: "I received your last message."
- Correct Use: "ROGER, proceeding as instructed, OUT."

The NATO phonetic alphabet is a recognized set of code words. These words are used to represent the 26 letters of the English alphabet. They are quite handy in noisy environments and can help avoid miscommunication of letters and numbers. Some of these alphabets include:

- A - Alpha
- B - Bravo
- C - Charlie
- D - Delta, and so on.

Here are a few practical examples of how these phrases are used in real-life:

Practical Example 3: Relaying a Grid Coordinate

- Incorrect Use: "Move to grid B R 6 2 9 4."
- Correct Use: "Move to grid Bravo Romeo Six Two Niner Four."

Practical Example 4: Identifying a Call Sign

- Incorrect Use: "This is team B2."
- Correct Use: "This is team Bravo Two."

Callsign in Tactical Communications

Alright so, the last thing we're going to talk about in this chapter is call signs. In tactical communication, call signs can help streamline communication and are also able to mask the identity of certain individuals from their adversaries. To keep it simple, a call sign is basically a coded identifier that's used instead of the name of a person or location.

These coded signs are often made of letters, numbers, or a combination of both and can be rotated to avoid detection and reduce risk. With that in mind, let's look at their strategic use and go over some practical examples and implementations.

Strategic Use of Callsigns

- **Concealment of Organizational Structure**: By assigning callsigns that do not directly correlate to the individual's rank or role within the organization, you can effectively mask the hierarchical structure from adversaries. This practice ensures that the importance or the specific skill set of an individual is not easily decipherable from their callsign.

- **Operational Security through Ambiguity**: Callsigns should be chosen to avoid any direct association with the individual's real name, role, or specific traits. This ambiguity protects the individual's identity and preserves operational integrity.

- **Frequent Rotation of Callsigns**: To counteract the efforts of intelligence collection and pattern analysis, it is critical to regularly change callsigns. This practice helps in evading detection and confusing adversarial tracking efforts.

Practical Examples

- **Example 1**: For a team conducting reconnaissance missions, instead of using call signs like "Recon1", "ReconLeader", which directly indicate their role and hierarchy, use abstract or unrelated call signs such as "Raven", "Cedar", "Delta2". These callsigns should be rotated periodically to prevent adversaries from developing an understanding of the team's composition and mission roles.

- **Example 2**: In an urban setting, where buildings or landmarks might serve as operational points of interest, assign random callsigns unrelated to the physical characteristics or importance of the location. For instance, a safe house located on Green Street could be designated "Watchtower" instead of anything suggestive of its purpose or location.

Implementation

- **Rotation Schedule**: Implement a predefined schedule for changing callsigns, such as after each mission, weekly, or any other interval deemed suitable based on operational tempo and threat assessment. Ensure all team members are briefed on the new call signs before they go into effect.

- **Secure Distribution**: Information regarding the assignment and rotation of callsigns should be distributed securely within the organization. Avoid electronic communication channels that are susceptible to interception. Utilize secure face-to-face briefings or encrypted messages for this purpose.

CHAPTER 15:
Guerrilla Tactics

Now that you've mastered Guerilla communications, we'll shift our focus to learning about how you can use Guerilla tactics with your Baofeng radio. To do this, we'll talk about things like Signal Operation Instructions (SOI), encrypted messages, developing a PACE plan, and more. There's a lot of ground to cover so let's begin.

The Critical Role of SOI

When it comes to tactical environments, secure communication is one thing you cannot leave to chance alone. SOI plays a critical role in securing communication during such an environment. What you need to know is that SOI includes everything from frequency usage, call signs, encryption codes, and timing for communication windows. It's a set of instructions which can be used by operational teams regardless of their locations or situation.

To better understand how SOI can be beneficial, think about a guerrilla group that's operating in a hostile territory. Now, in such circumstances, if they were to use radio transmission, they would end up drawing unwanted attention to themselves. That's not something you'd want when you're in enemy territory, do you?

In such cases, they can resort to an SOI that states that they should use low-power transmissions on pre-designated frequencies at specified times. But that's not all. The SOI would also state that the group should use coded messages to communicate. This would help them lower the risk of being detected and would allow them to keep information confidential.

.SOI Confidentiality and Management

92

You probably figured out that the SOI is a pretty important document. But what you need to know is how the document should be managed and kept secret. Why? Well, since it contains communication protocols, if it gets stolen, the damage can extend far beyond the operation. Believe it or not, it could compromise the entire framework.

What this means is that the SOI should be used and managed with confidentiality. Now, the important thing to understand here is that it may be an asset during an operation. However, as the operation ends, it can become a liability. So, the best practice, in such cases, is to destroy the document once it has served its purpose. Just so you get a better understanding, some of the key things that an SOI has include:

- **Frequency Hopping:** To communicate at predetermined times, switching frequencies according to a schedule known only to the team and command.

- **Call Signs:** The use of rotating call signs to identify individuals or units, changing daily to prevent pattern recognition by enemy forces.

- **Encryption:** Employing simple, manually encrypted codes for sensitive information, ensuring that even if transmissions are intercepted, the content remains secure.

- **Disposal:** Instructions for the immediate destruction of the SOI document using a burn bag or shredder after the mission's completion or upon compromise.

Developing and Implementing a Signals Operating Instructions (SOI) Chart

Developing and a Signals Operating Instructions (SOI) Chart

Now that you know what an SOI is, let's look at how you can develop it. An SOI outlines the communication protocols that are to be used by the team. It's divided into multiple components that include:

- **Frequency Table**: This section lists all the frequencies allocated for different purposes within the group, such as command, logistics, emergency, and so forth. Frequencies are assigned specific purposes to avoid confusion and ensure that communication lines remain clear and organized.

- **Callsigns**: Every individual, team, or unit within the operation is assigned a unique callsign. This practice masks the identity of the operators and units, contributing to operational security. Callsigns can be based on a predetermined theme or be randomly assigned.

- **Passwords**: To further authenticate communication and verify the identity of individuals or units initiating contact, passwords are implemented. These can change daily or per operational phase and are known only to the members within the operational structure.

- **SARNEG Grid**: The SARNEG grid is a simple yet effective method for encrypting and decrypting messages. It consists of a grid or chart that matches letters and numbers in a way that is unique to the group using it. This encryption is used for sensitive information that requires an additional layer of security beyond the inherent security measures of the radio system.

Out of all these components, we've already covered frequencies, callsigns, and passwords. So now, let's look at SARNEG. This is an encryption method that uses alphanumeric characters to represent numbers and often has a predefined grid or table. This grid is used to associate numbers from 0 to 9 with one or a combination of letters. For example, "1" might be encoded as "A," "2" as "B," and so on.

Imagine that you're **in** a situation where you and your search and rescue team need to communicate the following coordinates: Latitude 34.0522° N, Longitude -118.2437° W. To keep it simple, let's encrypt only the decimal part of the coordinates. Using a basic SARNEG-like system, here's what the encryption would look like:

- 0 = A
- 1 = B
- 2 = C
- 3 = D
- 4 = E
- 5 = F
- 6 = G
- 7 = H
- 8 = I
- 9 = J

Using this system, the coordinates would be encoded as follows:

- Latitude 34.0522° N becomes 34.FACB° N
- Longitude 118.2437° W becomes 118.CEDH° W

Based on this, the encrypted message would be "34.FACB° N 118.CEDH° W." You can now transmit the message to a recipient who would be able to decode it using the same system. Pretty cool, right!

Layers of SOI

The thing with an SOI is that each of its components can be used as layers when you implement it. This approach can help you enhance security and improve your communication methods. With that in mind, let's briefly go over the purpose, practical applications, and examples for each layer of the SOI/

Frequency Tables

- Definition and Purpose: Frequency tables provide a structured list of frequencies designated for specific operations or activities, ensuring team members tune into the correct frequency for coordinated communication.

- Practical Application
  - Compile a list associating each operational activity with a unique frequency, for instance, assigning 146.520 MHz for Patrol Operations and 446.000 MHz for Emergency communications.
  - Securely share this compiled list with all team members.
  - Regularly review and revise the frequency table to accommodate operational changes or respond to security threats.
- Example
  - For Patrol Operations, the assigned frequency is 146.520 MHz. In case of an emergency, switch to 446.000 MHz. Supply Drop Coordination occurs on 155.340 MHz.

Callsigns

Definition and Purpose: Callsigns are unique identifiers assigned to individuals or teams, replacing personal names to maintain privacy and operational security.

- Practical Application:
  - Callsigns can range from simple role-based identifiers, such as "Alpha Team," to thematic names like "Hawk" for a team leader, ensuring no direct link to personal identities.
  - Incorporate callsigns into the SOI, requiring all members to familiarize themselves with their own callsign and those of their teammates.
- Example:
  - The leader of Team Alpha is designated as "Falcon." The leader of Team Bravo goes by "Shadow," while the team's medic is known as "Lifesaver."

Passwords & SARNEG

- Definition and Purpose: Passwords authenticate the validity of transmissions, confirming the identity of communicators. SARNEG, a sophisticated encryption method, adds a layer of security to messages.
- Practical Application:
  - Allocate a daily or operational password for routine verification, changing this frequently to ensure security.
  - For encrypted communication, utilize SARNEG codes, a prearranged set of letters and numbers, to encode and decode messages, practicing their use to ensure efficiency.
- Example
  - The daily operational password is "Eagle Flight," refreshed every 24 hours. A SARNEG code, such as "5B9A2," is used for sensitive communications. For instance, the message "Proceed to rally point" would be encrypted to "8K3F7" using the designated SARNEG code.

Secure and Encrypted Messaging

For the next part of this chapter, we'll look at how you can ensure that all the messages you send are secure and encrypted. To do this, we'll look at in-person challenges, running passwords, a PACE plan, and more.

In-Person Challenge/Passwords

The challenge/password system is a time-tested method that you can use in various operational environments. It can help you confirm the identity of individuals or units as friendly forces. It's kind of like having a secret knock on your treehouse door. This system basically has two components: the "challenge" and the "response" or "password." Let's use an example that'll help you get a better understanding of how this method works.

Say that before a night operation, the team leader assigns a challenge/password combination. The challenge might be "Moon," and the response is "Stars." When a team member comes up to someone in the dark, they issue the challenge by saying "Moon." If the individual is friendly and knows the response, they reply with "Stars," confirming their identity. Pretty straightforward, right? Running Passwords for Distress Signals and Secure Radio Authentication

A running password is basically a code that keeps on changing. This code is used to authenticate radio transmissions and is a simple sequence of numbers or words. It's a great way to send distress signals and keep communications secure. Some examples of a running password may include:

- **Setup:** At the start of an operation, the running password is set as "Eagle1."
- **Evolution:** After each radio check or at predetermined intervals, the password increments (e.g., "Eagle2," "Eagle3," etc.).
- **Usage:** When initiating a radio transmission, the sender states the current running password. The receiver confirms by acknowledging with the next password in sequence ("Eagle2" receives, "Eagle3" acknowledges).

Implementing a PACE Plan for Effective Communications and Operational Security

PACE is basically a communication plan that refers to a Primary, Alternate, Contingency, and Emergency communication plan. It's useful when it comes to making communication efforts more secure and can be beneficial in tactical, rescue, or high-stakes environments. Let's go over all the layers of the PACE and along with an example and the equipment needed for each.

Primary Communications Plan

- The Primary aspect of the PACE plan involves the first-choice method for communication, route selection, or equipment use. For communications, this might be a specific VHF or UHF frequency deemed secure and reliable for the operation at hand.
- **Example:** For a search and rescue team, the primary frequency might be 146.520 MHz (2-meter amateur radio band), widely recognized for emergency communication among amateur radio operators.
- **Primary Route/Equipment:** The first-choice route for reaching an objective or the primary equipment set, such as GPS devices or satellite phones for navigation and communication.

Alternate Communications Plan

- In cases where the primary method is untenable or compromised, the Alternate plan provides a secondary option. This ensures the operation can continue with minimal disruption.
- **Example:** If interference or jamming affects the primary frequency, the team switches to an alternate UHF frequency, such as 446.000 MHz, pre-agreed and known to all members.

- **Alternate Route/Equipment**: A pre-planned secondary route or backup equipment, like traditional compasses and maps, if primary navigation tools fail.

Contingency Communications Plan

- Contingency plans address fewer probable scenarios that could significantly impact the operation. They are designed to recover from a partial failure of both primary and alternate plans.
- **Example**: In a scenario where radio communication is entirely compromised (e.g., due to jamming or equipment failure), the contingency plan might involve using encrypted messaging apps on smartphones, assuming cellular or satellite network availability.
- **Contingency Route/Equipment**: Routes and equipment planned for specific, less likely scenarios where primary and alternate routes are inaccessible or unsafe.

Emergency Communications Plan

- The Emergency component of the PACE Plan covers scenarios where immediate action is required to preserve life, such as personnel recovery or emergency signaling in distress situations. This layer often includes non-electronic signals or broadly recognized distress signals.
- **Example**: If an operative is injured and radio communication is not possible, emergency plans might include visual signals such as flares, smoke signals, or specific patterns of flashlight blinking known to search and rescue teams.
- **Emergency Route/Equipment**: Predefined emergency extraction points accessible by all team members, and personal locator beacons (PLBs) as last-resort equipment for indicating one's location to rescuers.

Now that you know all about each layer of the PACE plan, There are some things that you need to consider before implementing it. This includes:

1. **Clearly Define and Communicate**: Ensure that all team members understand the details of the PACE plan, including frequencies, routes, and equipment for each layer.
2. **Conduct Drills**: Regularly practice implementing the PACE plan under various simulated conditions to ensure smooth execution under stress.
3. **Use Codewords**: Implement simple, secure codewords for switching between layers of the plan without divulging details over potentially compromised channels.
4. **Review and Adapt**: After each operation or exercise, review the effectiveness of the PACE plan and make adjustments based on lessons learned and new intelligence.

Communication Windows for Operational Security

When it comes to communication windows, you need to know that using them is essential for operational security. Communication windows are basically predefined times where radio communication can be used. They are designed to reduce the risk of having an enemy predict when communications are likely to occur.

In high-risk environments these time slots can be used for check-ins, information exchange, and coordination efforts. But that's not all. When communication windows are implemented, you don't need

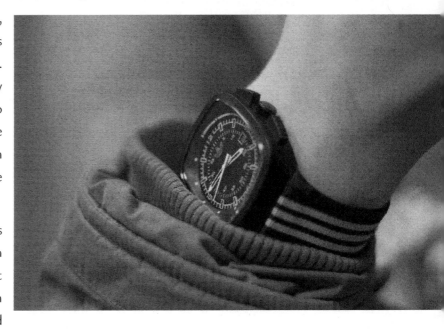

to use your Baofeng radio outside of them, meaning that you can switch off the device and make the battery last longer.

With that in mind, let's look at some practical implementations, examples, and security measures that'll help you develop and implement communication windows effectively.

Practical Implementation

- **Pre-Planning:** Before deploying or initiating operations, establish a clear communication plan that includes specific times for opening Communication Windows. These times should be agreed upon by all team members and understood to be the only periods for transmitting, barring emergency situations.

- **Synchronization of Timepieces:** Ensure that all team members' watches or timekeeping devices are precisely synchronized. A common failure in maintaining effective Communication Windows is discrepancies in team members' timekeeping.

- **Duration and Frequency:** Depending on the operational needs and security level, Communication Windows may range from a few minutes to longer periods but should remain as brief as practicable. The frequency of these windows can vary from several times a day to once every few days, based on mission requirements and the threat level of signal detection.

Example Scenario

- **Morning Check-In:** A team operating in a remote environment might establish a morning Communication Window at 0800 hours for 5 minutes. During this period, team members check in, report status, and receive any necessary updates or adjustments to the operation plan.

- **Evening Brief:** Another window might open at 2000 hours for 10 minutes, allowing for a more detailed exchange of information, such as intelligence gathered during the day and coordination of the next day's activities.

- **Emergency Protocol:** In addition to scheduled windows, an emergency protocol should be established, allowing for unscheduled communication if critical situations arise. This protocol should include a specific frequency and call sign that indicates an emergency transmission.

Security Measures

- **Randomization:** To further enhance security, consider slightly randomizing the timing of Communication Windows. For example, instead of always starting exactly on the hour, use a predetermined variance (e.g., +/- 10 minutes) that team members can calculate based on a shared secret or pattern.

- **Silent Monitoring:** Outside of these windows, radios can be set to a receive-only mode to monitor emergency calls or gather intelligence without transmitting a detectable signal.

- **Frequency Hopping:** If operationally feasible, change frequencies following a pre-established pattern for each Communication Window to reduce the risk of adversaries tracking or jamming frequencies.

Reporting in Tactical Operations

Another thing that's very important in tactical communication is reporting. When it comes to reporting protocols, there are a lot of different options like SALUTE Reports, SALT Reports, and the 9-LINE MEDEVAC. Now, we don't need a very detailed understanding here, so I'll just briefly go over what each of these reports are and share some examples along the way.

SALUTE Reports

First up we have the SALUTE report. SALUTE stands for Size, Activity, Location, Uniform, Time, and Equipment. If you were thinking that it's just a fancy name thing again because this is a system that ensures all relevant information about the enemy's activities is covered. Here's an example of how the information in such a report would look.

- **Size:** Approximately 15 personnel.
- **Activity:** Moving eastward along the ridge line, possibly scouting or searching for something.
- **Location:** Three kilometers north of Hill 455.
- **Uniform:** Camouflage uniforms, similar to local forest terrain.
- **Time:** Observed at 0930 hours.
- **Equipment:** Light arms observed, AK-47s predominantly, with two personnel carrying RPGs.

SALT Reports

The SALT report is used as a supplement to the SALUTE report. It provides updates about the activities that were observed and helps maintain the flow of information. It follows the same acronym and focuses on Size, Activity, Location, and Time. Some of the things these reports might include are:

- **Size:** The group has increased to 20 personnel.
- **Activity:** Now setting up a temporary encampment.
- **Location:** Still near Hill 455, but has moved slightly south towards the creek.

- **Time:** Update as of 1100 hours.

9-LINE MEDEVAC

The 9-LINE MEDEVAC is used for requesting medical evacuation. It has specific details that help ensure timely and effective medical care. The nine lines of the report cover location, radio frequency, patient count by precedence, special equipment, patient type, security at pick-up site, HLZ marking, nationality, and NBC conditions. Here's an example of how it would look in real-life. Here's how the content of the report would look like:

1. **Grid:** 12345678 (exact pick-up location).
2. **Radio Frequency and Call Sign:** 150.350 MHz, Call Sign "Medic One".
3. **Patient Count:** 2 Urgent, 1 Priority.
4. **Special Equipment Needed:** None.
5. **Number of Patients:** Litter: 2, Ambulatory: 1.
6. **Security at Pick-up Site:** Secure.
7. **HLZ Marking:** Red smoke.
8. **Nationality and Status:** US Military.
9. **NBC Contamination:** None.

Clandestine Reports

These reports are essential for guerrilla forces that are operating in hostile territory. They can be used for communication between active field units and their command structures. Before we wind up this chapter, let's dig into the details of the different types of clandestine reports.

ANGUS: Initial Entry Report

The ANGUS report is used to record the start of an operation within enemy territory. It provides information for future actions and decisions by highlighting what the team's initial status and surroundings are.

Tactical Considerations:

- **Immediate Environment Assessment:** Quick and accurate portrayal of the operational environment highlighting any immediate threats or opportunities.
- **Initial Stealth:** Emphasizes the importance of remaining undetected upon entry, ensuring the report does not compromise the team's position.

- **Preparation for Future Reports:** Establishes a baseline for future communications and operations, detailing initial plans and intended movements.

Examples

- **Code Communication:** "Eagle Nest established at Dawn. Quiet as the forest. Hawks' unseen. Proceeding to River's Bend for a closer look. Will sing at dusk."

- **Interpretation:** The team has safely infiltrated the operational area undetected at dawn. No enemy forces (hawks) were observed. The team plans to move towards a strategic vantage point (River's Bend) for reconnaissance and will report back by the evening (sing at dusk).

BORIS: In-depth Intelligence Report

BORIS reports contain detailed information on enemy activities. This information can include their strengths, weaknesses, and potential targets, supporting strategic planning and targeted operations.

Tactical Considerations:

- **Long-term Surveillance:** Results from prolonged observation efforts, requiring stealth and patience to avoid detection while gathering detailed intel.

- **Analysis and Recommendations:** Beyond mere observation, BORIS reports include analysis of gathered intel and propose actionable recommendations for command.

- **Resource Allocation:** Helps command allocate resources effectively, targeting vulnerabilities identified through detailed intelligence.

Examples

- **Code Communication:** "The Orchard bears fruit; worms are scarce but the soil is rich. Farmer Jones visits twice a day, unaware of the hidden roots. Awaiting the harvest moon."

- **Interpretation:** The area is rich in targets (fruitful), with minimal enemy activity (worms are scarce). An enemy commander (Farmer Jones) unknowingly passes by the team's position frequently. The team is gathering more intel and waiting for an optimal time to strike (harvest moon).

CYRIL: Situation Report

CYRIL reports are used for getting regular updates on the team's status. These updates can be about changes in position, completed activities, encountered challenges, and next steps.

Tactical Considerations:

- **Operational Continuity:** Ensures the higher command remains informed of the ground situation, facilitating timely support and adjustments to plans.

- **Adaptability:** Reflects the team's ability to adapt to changing situations, crucial in dynamic and potentially hostile environments.

- **Mission Progress Tracking:** Offers a methodical way to track the progress of specific operations or missions, detailing successes and setbacks.

Examples

- **Code Communication:** "Current camp under the Old Oak remains undisturbed. The trail was windy but passable. Encountered a fallen tree, path redirected. Next sunrise, we move north."

- **Interpretation:** The team's current location is secure and they've remained undetected. They've encountered minor obstacles but have adapted their route. Plans are in place to relocate northward at the next opportunity.

CRACK: Battle Damage Assessment

CRACK reports are used to assess the effectiveness and impact of the operation once it is completed. They evaluate both the tactical success and strategic implications.

Tactical Considerations:

- **Objective Evaluation:** Measures the success of an operation against its objectives, critically assessing both the execution and outcome.

- **Future Planning:** Insights from CRACK reports inform future operations, highlighting effective tactics and areas for improvement.

- **Enemy Response Anticipation:** Evaluates the potential impact on enemy movements and strategy, anticipating and preparing for retaliatory actions or adjustments.

Examples

- **Code Communication:** "The storm passed; the dam shows cracks, but the reservoir holds. The scarecrow lost its hat. Awaiting the next rain to test the waters."

- **Interpretation:** An operation (storm) was carried out causing some damage to the enemy's infrastructure (dam shows cracks), but their larger capabilities (reservoir) remain intact. A minor enemy asset (scarecrow) was eliminated. The team is evaluating the impact and planning the next move.

UNDER: Cache Report

UNDER reports provide information about the creation and specifics of supply caches. This ensures that vital resources are securely stored and can be reliably accessed by guerrilla forces.

Tactical Considerations:

- **Operational Sustainment:** Caches represent a critical resupply point for ongoing operations, enabling extended presence in operational areas.

- **Security and Concealment:** Emphasizes the importance of hiding and securing caches to prevent discovery by enemy forces or non-combatants.

- **Strategic Resource Placement:** Involves strategic thinking about the placement of caches to support future operations, escape routes, and fallback positions.

Examples

- **Code Communication:** "A new hive was found beneath the Willow's shadow, guarded by the moon's gaze. It holds honey and stingers for the swarm. Marked by the sign of the Wren."

- **Interpretation:** A supply cache (hive) has been established in a secure location (beneath the Willow's shadow), containing resources and weapons (honey and stingers) for the team (swarm). It's discreetly marked for identification (sign of the Wren) and is under natural surveillance (moon's gaze).

CHAPTER 16:
Secure Communications in Diverse Operations

We have come quite a long way from being a newbie to learning all about the Baofeng radio and how it can be used. In this next chapter, we're going to cover how you can make communication efforts secure in diverse environments. To do this, we'll look at security and discipline aimed at making communication more effective. But that's not all. We'll also look at stealth communication techniques and dive deeper into things like sustainment and clandestine communications we covered earlier. Let's begin!

Strategic Radio Use

This one might seem like something that's quite common. But it's one that most people don't know about or can' implement. The strategic use of radio devices in tactical environments is necessary for operational security. Doing so can involve limiting access and developing communication and training protocols. Let's look at each of these in more detail.

Limiting Radio Access

- **Leadership Communication:** Designate radios primarily for leaders who require real-time information to make decisions and coordinate actions. This ensures that strategic and tactical directives are communicated efficiently without cluttering the airwaves.

- **Specialized Roles:** Assign radios to team members with specialized communication duties. This includes scouts, forward observers, or those tasked with liaising with other groups or units. Their use of radio is focused on their specific operational roles.

Communication Discipline

- **Predefined Protocols:** Establish clear communication protocols that define when and how to use the radio. This includes using codewords, predetermined frequencies, and specific times for communication to reduce the likelihood of interception and exploitation by adversaries.

- **Minimal Traffic:** Encourage minimal radio traffic by emphasizing the importance of concise, relevant, and pre-scripted messages. This reduces the chance of electronic detection and ensures that critical information is not lost in unnecessary communication.

- **Operational Security:** Regularly remind all team members of the importance of OPSEC. Casual or non-essential use of radios can inadvertently reveal operational details or patterns that can be exploited by opponents.

Training and Enforcement

- **Regular Drills:** Conduct regular communication drills to reinforce discipline and ensure familiarity with protocols. This helps in acclimatizing team members to operate under constraints and understand the gravity of secure communication.

- **Enforcement Measures:** Implement measures to monitor and enforce communication discipline. This could include checks on radio usage, debriefs on communication effectiveness, and corrective actions for breaches of protocol.

Sustainment Communications

Remember what we learned about sustainment communication? Great! Now, it's time to build on that knowledge and get a deeper understanding. The most important thing to remember about this type of communication is that it's used to build resilience in environments where traditional communication methods are not working. Let's look at the importance, examples, and applications of sustainment communication.

Importance in Crisis Situations

- **Emergency Response:** In the aftermath of natural disasters, when standard communication networks may be down, sustainment communications provide a vital lifeline, enabling effective coordination of relief efforts.

- **Oppressive Regimes:** Governments or hostile forces may attempt to suppress dissent by shutting down communication networks. Sustainment communications offer a way to circumvent censorship, keeping channels open for organizing, reporting abuses, and spreading uncensored information.

- **Guerrilla Coordination:** For guerrilla groups or resistance movements, these communication networks are essential for coordinating activities, sharing intelligence, and maintaining operational security against a better equipped adversary.

Historical and Modern Examples

- **Radio Rebelde:** During the Cuban revolution, Radio Rebelde was set up by Che Guevara and his comrades to broadcast news and revolutionary propaganda across the island, playing a key role in their information warfare strategy. This early example demonstrates the power of radio to mobilize support and disseminate critical information even in resource-constrained conditions.

- **Modern Guerrilla Movements:** In contemporary contexts, guerrilla groups and activists can harness sustainment communications to create decentralized networks using readily available technology like the Baofeng radios. This enables them to remain agile and resilient, even when facing surveillance or jamming efforts from state level actors.

Practical Applications

- **Radio Networks:** Setting up a network of handheld radios (such as Baofeng UV-5R models) among a community or resistance movement can provide a robust and flexible communication system. These networks can support everything from coordinating supply drops and medical aid to planning strategic movements and sharing intelligence.

- **Pirate Radio Stations:** Low-power FM transmitters can be used to establish pirate radio stations, broadcasting critical information, instructions, or propaganda to a wider audience. This can be particularly effective in rallying support, spreading awareness of issues, or countering disinformation campaigns by hostile forces.

- **Digital and Encrypted Communications:** For more technologically advanced scenarios, digital modes and encryption can enhance the security and efficacy of sustainment communications, making it more difficult for adversaries to intercept or disrupt messages.

Establishing Sustainment Communications

Alright now that you're aware of what sustainment communication is, you can just follow the steps below to implement it.

1. **Identify Needs and Resources:** Assess the communication needs of your group or community and the resources available. This includes hardware (radios, antennas, power sources) and knowledge (operators familiar with radio operation and encryption techniques).

2. **Train and Educate:** Provide training on the use and maintenance of the communication equipment, as well as basic operational security practices to safeguard the network.

3. **Plan and Coordinate:** Develop a communication plan that outlines protocols, frequencies, call signs, and encryption keys (if applicable) to ensure everyone is on the same page and can communicate effectively when needed.

4. **Test and Drill:** Regularly test the communication system through drills and exercises to identify any weaknesses and familiarize users with the equipment and procedures under various scenarios.

Coordination and Stealth

When it comes to tactical communication, you must be aware of the fact that both coordination and stealth are highly important. Let's use an example to explain tactical communication further just so you have a clear understanding. Let's say that you're coordinating an ambush. During the ambush, one of the team members spots the target and uses a code word that's transmitted on a low power setting to alert the main force.

As soon as the alert comes in the TOC is aware of the fact that the target is in the strike range. The TOC then signals the different teams telling them to get ready in their positions. As the target approaches, another signal is sent to initiate the ambush.

And that is how tactical communication's work! Some of the factors that you must be aware of to master tactical communication include:

- **Brevity is Key:** Keep transmissions short and to the point. The longer you transmit, the higher the chance of detection and interception.

- **Codewords are Crucial:** Develop and use codewords for common phrases and locations. This not only speeds up communication but also adds a layer of security.

- **Low Power Settings:** Utilize the lowest power setting that allows for clear communication within your team. This minimizes the signal's detectable range, reducing the likelihood of interception.

- **Strategic Positioning:** Avoid transmitting from high points or open areas where your signal has the farthest reach. Position yourself in valleys or just below ridge lines to use the terrain to your advantage, making direction finding by the enemy more difficult.

In addition to these factors, you must also know that a Tactical Operations Center (TOC) is where communications and operations are managed. When developing a TOC setup, you need to consider:

- **Discreet Location:** Choose a location that's defensible and hard to pinpoint. Avoid areas that are too isolated or too exposed.

- **Omni-directional Antennas:** For broad communication needs, omni-directional antennas can provide the necessary range while maintaining a low profile.

- **Digital Security:** Ensure all digital communications are encrypted and employ counter-surveillance measures to detect any attempts at interception.

- **Communication Discipline:** Train all operators in strict communication security (COMSEC) practices to avoid leaks of sensitive information.

Communication Techniques for Secure Operations

When it comes to conducting secure operations, clandestine communications are one of the most feasible options. know that we've already covered this earlier, so we'll skip the basics and dive right into some of the factors that you need to consider when implementing such communication protocols. These factors include:

- **Directional Antennas:** Utilize directional antennas such as Yagi antennas for focused transmission. This limit the spread of the signal to a narrow path, significantly reducing the likelihood of interception by adversaries These antennas should be aimed precisely at the predetermined reception point for maximum efficiency and security.

- **Pre-Planned Transmission Sites:** Select transmission sites carefully in advance. These sites should have a clear line of sight to the receiving station while utilizing natural terrain features to obscure the signal from unintended directions. Transmission sites should be located at least 1000 meters away from sensitive locations such a operational bases or hide sites to protect against direction-finding efforts by hostile forces.

- **Avoiding Sensitive Locations:** Never transmit directly from or near your operational base, hide sites, or any sensitive locations. The act of transmitting can make these sites vulnerable to enemy surveillance and direction finding technologies. Always assume that the enemy is monitoring and capable of triangulating your transmission source.

- **Using Terrain to Your Advantage:** Whenever possible, leverage the natural terrain to mask your radio signals Transmitting from low areas or behind terrain features can help conceal the signal's origin and make it more challenging for adversaries to detect or locate. This technique is particularly effective in hilly or mountainou areas where the terrain can naturally deflect or absorb radio waves.

- **Minimal Transmission Time:** Keep transmissions brief and to the point. The longer you transmit, the greate the chance of detection. Plan your messages in advance to ensure they are concise and contain all necessary information without unnecessary chatter.

- **Regularly Changing Frequencies:** Avoid using the same frequency for extended periods. Regularly changing frequencies can complicate enemy efforts to monitor and decrypt your communications. This should be a par of your pre-planned communication strategy, with all team members aware of the frequency change schedules

- **Secure Encoding and Encryption:** Use advanced encryption methods and secure encoding techniques to protect the content of your communications. Even if an adversary can detect the transmission, encryption ensures that the message remains confidential and undecipherable.

Now that you're well aware of what you need to consider when implementing clandestine communication protocols just follow the steps below to implement them effectively.

1. **Set Up:** Prior to an operation, set up directional antennas at your chosen transmission site, ensuring they are camouflaged and pointing towards your intended receiver. Now you need to test the setup to confirm that communication is clear and secure.

2. **Operation:** During operations, communicate using pre-agreed codes and encryption, transmitting only a scheduled times or as absolutely necessary. It's important to keep the transmissions brief and use low powe settings to further reduce detection risk.

3. **Mobility:** After transmitting, quickly relocate to prevent any potential triangulation of your position. An important thing worth remembering is that you should always have multiple pre-planned exit routes from your transmission site.

CHAPTER 17:
Encoding and Encryption Strategies

When it comes to Guerrilla situations, one of the most critical things in such operations is secure communication. To increase security, you can implement encoding and encryption protocols. In this chapter, we'll focus on learning the difference between the two and how they can be combined. Then, we'll look at different types of communication in an operational environment and cover things like Trigram encryption and more.

Encoding vs. Encryption

Both encoding and encryption are keyways to safeguard information that are used to enhance the security of information transmission. Let's look at each one alongside its practical use and implementation.

Encoding

- Encoding is the process of transforming information into a specific code or format that can be easily understood by someone with the necessary knowledge or key to decode it. It's practical use and implementations include:

    - **Practical Use:** For instance, you might encode a message using a simple alphanumeric substitution system where each letter is replaced by a corresponding number (A=1, B=2, etc.). To encode the word "HELP," you would transmit "8-5-12-16."

    - **Implementation:** Develop a pre-agreed encoding system within your team. Ensure all members understand and can swiftly apply the system. Use this for non-critical communications that benefit from quick decoding by team members while maintaining a basic level of obscurity.

Encryption

- Encryption takes the security of information one step further. It is used to scramble the message in a way that it cannot be understood by those for whom it's not intended. The practical use and implementation of encryption is mentioned below.

  - **Practical Use:** Utilizing a symmetric encryption algorithm, such as AES (Advanced Encryption Standard), you could encrypt the message "MEET AT DAWN" so it appears as a random string of characters, e.g., "G7zhJQm4Po9aV...". Only those with the correct decryption key can revert it to its original form.

  - **Implementation:** Choose an encryption method that suits your operational needs and ensure all communication devices or software used by the team support this encryption. Regularly update and securely distribute encryption keys among team members.

Combining Encoding and Encryption for Maximum Security

Both encoding and encryption can be combined to get the maximum security possible. Implementing such a communication technique means that those who want to hear the message would first need to decrypt it and then decode it.

What this means is that you would first need to encode a message ("MEET AT DAWN" to "13-5-5-20 1-20 4-1-23-14") and then encrypt the encoded message into something that's impossible to understand. This makes it challenging for unauthorized individuals to gain access to sensitive information.

To combine both these encoding and encryption you need to:

1. Develop and agree upon an encoding scheme within your team.
2. Choose a robust encryption method compatible with your communication tools.
3. Train all team members in both encoding and decrypting messages to ensure smooth operation.
4. Regularly review and update your encoding schemes and encryption methods to counter potential security threats.

Strategies Across Different Operational Contexts

Remember the communications types we learned earlier? Well, we're not done with that just yet. Sustainment, tactical, and clandestine communications can be used in different operations contexts. Let's dive into the details of each one to deeper understanding.

Sustainment Communications for Balancing Efficiency and Security

Two keys that you need to consider when using sustainment communication include:

- Utilization of Digital Modes: For more efficient data transmission, leveraging digital communication modes such as andFLmsg provides a robust platform. This system allows for rapid, keyboard-to-keyboard data exchange, significantly condensing transmission time and reducing the potential for human error during message relay.

- Security Measures: Despite the lower COMSEC requirements, it's vital to employ basic security measures. This includes the use of encoding for sensitive information, careful frequency selection to avoid interception, and the regular review of communication plans to adapt to any emerging threats or vulnerabilities.

Tactical Communications for Precision and Stealth

When it comes to tactical communication, you need to ensure:

- Short and Secure Transmissions: Communication should be concise, encrypted (where possible), and transmitted for the shortest time necessary to convey the message. This minimizes the window for enemy forces to detect and triangulate the signal source.

- Low Power and Minimal Antenna Usage: To further reduce the likelihood of detection, transmissions should be conducted on the lowest power setting that still allows for clear communication between parties. Additionally, the use of minimal antenna setups limits the signal's range, confining it to the operational area and reducing the chance of interception by outside entities.

- Leadership Control of Communication Equipment: To maintain discipline and operational security, radio equipment should primarily be managed and operated by leadership or designated communications specialists. This centralizes control over the transmission process and ensures that communication protocols are strictly followed.

Clandestine Communications for Maximizing Operational Security

As far as clandestine communication is concerned, you can use:

- Directional Antennas for Focused Transmission: Utilize directional antennas to concentrate the radio signal towards the intended receiver, significantly limiting the risk of interception by adversaries.

- Pre-Planned Transmission Sites: Operatives must select transmission sites strategically, ensuring they are well away from sensitive locations or operational bases. This planning includes considerations for natural terrain features that can mask the origin of the transmission, complicating enemy efforts to locate the signal source.

- Rigorous COMSEC Practices: Beyond the use of encrypted digital modes, clandestine operations may also employ traditional methods like one-time pads for encryption or pre-arranged code words. The selection of transmission frequencies should be unpredictable, with operatives ready to change them to avoid pattern recognition by hostile forces.

Trigram Encryption for Secure Communication

When it comes to encrypting communications, using the Trigram methods is a great option. It uses a set of three letters that represent a word or a phrase. If you want to use this method you'll need to convert sentences into a series of trigrams. This method adds a critical layer of security as a key is needed to decrypt the messages.

To implement trigram encryption, all you have to do is follow the steps below:

Step 1: Generating Trigram Keys

- Create a comprehensive list of words and phrases frequently used in your communications.

- Assign a unique trigram to each word or phrase. Ensure there are no duplicates to avoid confusion.

Step 2: Encrypting Messages

- Write out your message as you normally would.

- Replace each word or phrase with its corresponding trigram from your key.

- If a word in your message doesn't have a pre-assigned trigram, use a generic method to encode it or update your key to include it.

Step 3: Transmitting Encrypted Messages

- Once your message is fully encrypted into a string of trigrams, it's ready for transmission. The recipient, using the same key, can decrypt the message back into its original form.

Here's a neat little example I made for you:

Suppose your trigram key includes the following assignments:

- HEL = Help
- EXF = Extraction
- LOC = Location
- IMM = Immediate
- DNG = Danger

In this example the message is "Immediate extraction needed at location Bravo," and when encrypted with Trigrams it becomes "IMM EXF NDD AT LOC BRV."

When someone receives the message, they can refer to the shared trigram key to translate each one back into its original words. When using trigrams, some of the key things you need to remember include:

- Regularly update and rotate trigram keys to maintain security.
- Limit the distribution of keys to essential personnel only.
- Practice encrypting and decrypting messages to become fluent in using trigrams efficiently.

One Time Pad Encryption

The One Time Pad (OTP) encryption is another secure communication method that can be used to encrypt messages mathematically. The key principles for OTP messages include:

- Unique Key for Each Message: The essence of OTP's security is the use of a randomly generated key that is as long as the message itself. This key is used only once and then discarded.
- Secure Key Exchange: The sender and receiver must both possess the same key, distributed through secure channels. The entire security of the OTP system hinges on the key remaining secret.
- Concise and Clear Messaging: Before encryption, messages should be clear and concise to ease the encryption process. Each character in the message is then assigned a numeric value.
- Encryption Process: Encrypting the message involves a simple arithmetic operation (typically addition) of the message's numeric code with the numeric code of the key.
- Transmission of Encrypted Message: The encrypted message, now a series of seemingly random numbers or characters, can be safely transmitted without the risk of decryption by unintended parties.
- Decryption by the Receiver: Using the identical key, the receiver performs the inverse arithmetic operation to decrypt the message and convert the numeric codes back to text.

Let's say you want to encrypt the message "HELP" using the One-Time Pad (OTP) method where each letter is assigned a numerical value (A=1, B=2, ..., Z=26), the message would be converted as follows:

- H = 8
- E = 5
- L = 12
- P = 16

So, "HELP" would be 8-5-12-16 numerically.

For practical application, let's say our OTP key for this message is "739485." To encrypt:

1. H (8) with the first digit of OTP (7) would be encrypted by adding the two (15).
2. E (5) with the second digit of OTP (3) becomes 8.
3. L (12) with the third digit of OTP (9) turns into 21.
4. P (16) with the fourth digit of OTP (4) results in 20.

What this means is that the encrypted message would be "15-8-21-20." To decrypt this message, the recipient, who has the same OTP key, would subtract the key from the encrypted message to retrieve the original numerical message. To make sure that you are using the OTP effectively, you need to consider the following.

- Key Security: Absolute security of the key is paramount. If the key is compromised, the encryption's integrity is void.
- Single Use of Keys: Never reuse keys. A new, randomly generated key is required for each message to maintain security.
- Accuracy is Critical: Precision in encryption, transmission, and decryption is vital. Any errors can make the message undecipherable.
- Destruction of Keys Post-Use: To prevent any potential security breaches, both the sender and receiver should destroy the used key securely.

CHAPTER 18:
Baofeng in Natural Catastrophes

Natural disasters can strike at any given moment and are unavoidable. If severe enough these disasters tend to push the limits of our resilience and ability to survive. During such circumstances, communication is one factor that can help deal with the aftermath of such events.

Severe natural disasters often lead to traditional means of communication being unavailable. If stranded somewhere during such scenarios finding a way out without being able to communicate can be challenging. However, this can be avoided if you know how to use your Baofeng radio during such events.

Preparedness Measures for Natural Disasters

Natural disasters, be it a hurricane, floods, wildfire, earthquake, snowstorm, are all hard to predict. Given this, you need to be proactive when it comes to preparing for such events and one of the many, some might argue the most important, thing to do as a proactive measure is to learn how you can make your Baofeng radio a part of your disaster preparation plan. This can seem like a bit tough, but it won't be if you follow the steps mentioned below:

Understanding the Disaster Profile:

- Each natural disaster has its unique challenges. Begin by understanding the types of natural disasters most likely to occur in your area. This knowledge will guide you in tailoring your Baofeng radio's use, from programming specific emergency frequencies to preparing for the environmental conditions likely to be encountered.

Programming Essential Frequencies:

Prioritize programming your radio with frequencies that are critical during disasters. This includes local emergency services, NOAA Weather Radio (NWR) frequencies for real-time weather updates, and amateur radio emergency service (ARES) frequencies, where licensed ham operators provide emergency communication support.

Creating a Communication Plan:

Develop a comprehensive communication plan that includes contact information for emergency services, family, and friends. This plan should outline how and when to use your Baofeng radio to communicate during a disaster. Include alternative communication methods should traditional networks fail. Ensure all family members are familiar with the plan and know how to operate the Baofeng radio.

Regular Maintenance and Testing:

Keep your Baofeng radio in optimal condition by performing regular maintenance and testing. This includes checking the battery life, ensuring all programmed frequencies are current, and verifying that the radio remains waterproof and dustproof, especially in anticipation of harsh environmental conditions.

Emergency Power Solutions:

Since power outages are common during natural disasters, having alternative power sources for your Baofeng radio is crucial. Solar chargers, hand-crank generators, and spare batteries can ensure your radio remains operational when you need it most.

Training and Drills:

Knowledge and preparedness go hand in hand. Participate in community emergency response team (CERT) training or amateur radio emergency service (ARES) drills to hone your skills in using the Baofeng radio under emergency conditions. These exercises simulate real-world scenarios, teaching you how to communicate effectively, manage power resources, and access critical information during disasters.

Waterproofing and Protection:

Natural disasters often expose your radio to water, mud, and debris. Waterproof cases or bags specifically designed for electronics can protect your Baofeng radio from the elements. Additionally, consider floatation devices for your radio if you're in a flood-prone area, ensuring it doesn't get lost in water.

Portable Antenna Solutions:

Enhancing your radio's capability with a portable, high-gain antenna can be a game-changer in a natural disaster. A better antenna can increase your transmission range, crucial for reaching out to emergency services or connecting with community response teams over greater distances.

Documentation and Manuals:

Finally, include a waterproof copy of your Baofeng radio's manual and your communication plan with your emergency kit. In the stress of a disaster, having a quick reference can be invaluable, especially for those less familiar with the radio's operation.

Real-Time Disaster Information Acquisition

Having access to timely and accurate information during such events is essential. But, if traditional communication and information sharing methods are working, you can always rely on your Baofeng radio.

It's possible that you might not be aware of what to do when it comes to configuring your Baofeng radio for real-time disaster information. But there's no need to go into panic mode. Some of the key things you can do to get such information include:

Accessing NOAA Weather Radio (NWR) and Emergency Alerts:

By programming your Baofeng to receive NOAA Weather Radio broadcasts, you gain access to a continuous source of weather updates and emergency alerts. This real-time information is crucial for understanding the scope of the disaster, potential changes in conditions, and recommended actions for safety. The capability to monitor multiple channels ensures that you can stay informed about developments both locally and in broader areas.

Amateur Radio Networks:

The amateur radio community often becomes a primary source of information during disasters, particularly when traditional communication networks are down. By connecting with local amateur radio operators or networks, you can receive updates on the disaster's impact, learn about areas in need of assistance, and get insights into the status of recovery efforts. This community-driven information is invaluable for assessing situations and planning your response.

Scanning Local Emergency Frequencies:

Setting your Baofeng to scan mode allows you to monitor local emergency frequencies, providing insights into the ongoing response from law enforcement, fire departments, and emergency medical services. Listening to these communications can offer immediate information on evacuation orders, road closures, and areas of severe impact.

Rescue and Relief Operations

Another important thing that you need to be aware of during natural disasters is how to coordinate relief and rescue operations. The operation can help you get or provide necessary aid during natural disasters and, in some cases, this aid is what can be the difference maker. Some of keyways you can use you're Baofeng radio for relief and rescue include:

Establishing Communication Hubs:

In the aftermath of a disaster, establishing a central communication hub using Baofeng radios can facilitate the coordination of rescue and relief efforts. This hub serves as a point of contact for volunteers, local authorities, and affected individuals, ensuring that information is accurately relayed and resources are efficiently allocated.

Volunteer Networks and Mobilization:

The Baofeng radio enables the rapid mobilization of volunteer networks. By creating dedicated channels for different volunteer groups—search and rescue teams, medical personnel, supply distribution teams—you can streamline communication, coordinate logistics, and deploy resources where they are needed most. The clarity and immediacy of radio communication ensures that volunteers receive real-time instructions and can report back on their progress and challenges.

Interoperability with Other Agencies:

One of the challenges in disaster response is ensuring interoperability among various responding agencies. The Baofeng radio, with its flexibility in programming and ease of use, can be adapted for cross-agency communication. By establishing agreed-upon frequencies for interagency communication, you can enhance collaboration, share critical information, and avoid duplication of efforts.

Information Relay to Affected Populations:

Beyond coordination efforts, the Baofeng radio can be used to relay information directly to affected populations. Broadcasts can include safety instructions, locations of shelters, availability of medical aid, and updates on the status of recovery efforts. This direct line of communication ensures that individuals have the information they need to navigate the aftermath of the disaster safely.

Documentation and After-Action Reviews:

Finally, the use of Baofeng radios in coordinating rescue and relief operations provides a record of communications that can be invaluable for after-action reviews. Analyzing these communications can offer insights into the effectiveness of response efforts, identify areas for improvement, and inform planning for future disasters.

CHAPTER 19:
Communication in Urban Survival Scenarios

Urban environments are known for having a dense infrastructure and complex landscapes. This is a problem for most people as they live their day-to-day lives. However, it poses severe communication challenges in crisis situations. In this chapter, we're going to look at how you can use your Baofeng radio to enhance your communication effectiveness in such an environment.

Overcoming Urban Obstacles

High-rise buildings, underground roads, electric noise from millions of devices, all of this is synonymous with urban cities. Where all of these things facilitate a modern lifestyle, they can create communication problems. However, there are ways you can overcome these obstacles using your Baofeng radio. To do that, you need to focus on:

Understanding Signal Propagation:

The first step in overcoming urban communication challenges is understanding how radio signals propagate in dense environments. High-frequency signals, while effective in open areas, may be absorbed or reflected by buildings and other structures in urban settings. Utilizing the VHF and UHF bands of your Baofeng radio, you can find a balance between range and penetration, optimizing communication within the urban fabric.

Strategic Antenna Placement:

The effectiveness of your Baofeng radio can be significantly enhanced by strategic antenna placement. In an urban survival scenario, consider attaching your antenna to the exterior of a building or placing it on high ground to extend the range of transmission and reception. Employing a longer, aftermarket antenna can also improve your radio's ability to pick up and send signals amidst urban obstacles.

Utilizing Repeaters:

Many urban areas are equipped with amateur radio repeaters that can extend the reach of your Baofeng radio far beyond its line-of-sight capabilities. By programming these repeaters into your radio, you can leverage the existing infrastructure to maintain communication across the city, even when direct communication is not possible due to obstructions.

Building-to-Building Communication:

In situations where widespread communication systems are down, establishing direct, building-to-building communication networks can be vital. The Baofeng radio's ability to operate on both VHF and UHF frequencies allows for flexible communication strategies, adapting to the specific needs and challenges of the urban environment.

Improvised Repeater Networks:

In the absence of formal repeater infrastructure or in cases where it has been compromised, it is possible to set up improvised repeater stations using additional Baofeng radios. This guerrilla approach to communication infrastructure can be instrumental in coordinating rescue operations, distributing resources, and sharing information during prolonged urban survival scenarios.

Digital Modes for Urban Communication:

Incorporating digital modes of communication, such as DMR (Digital Mobile Radio), can overcome some of the limitations posed by urban clutter. Though Baofeng radios are primarily analog devices, certain models and modifications allow for the transmission of digital signals, which are more resilient to interference and can carry more information with less power.

Navigating Electronic Interference:

Urban areas are saturated with electronic devices that can cause interference, including Wi-Fi routers, microwave ovens, and other radios. To mitigate this, focus on using frequencies that are less crowded and employ CTCSS (Continuous Tone-Coded Squelch System) or DCS (Digital Coded Squelch) tones to filter out unwanted noise, ensuring clearer communication.

Power Management in Urban Settings:

Effective power management becomes crucial in urban survival scenarios, where recharging opportunities may be limited. Utilizing power-saving features on your Baofeng radio, carrying spare batteries, or employing solar-powered chargers can ensure that your communication lifeline remains active when you need it most.

Community Communication Networks:

Establishing and participating in community communication networks can provide a critical support structure during urban disasters. By coordinating with local emergency response teams, neighborhood watches, and other civic groups, you can create a resilient web of communication that enhances safety and resource distribution for everyone involved.

Urban Communication Infrastructure

When it comes to communicating during a disaster, one of the most important things to do is develop a communication infrastructure. To create such an infrastructure, we'll use the Baofeng radio. This can be done in several different ways that include:

Community Radio Networks:

The foundation of an urban communication infrastructure can be built through the formation of community radio networks. These networks, organized by blocks or neighborhoods, can utilize Baofeng radios to create a mesh of communicators. Each household or participant with a radio acts as a node within this network, capable of passing along messages, sharing updates, and coordinating local efforts.

Central Communication Hubs:

Designating central communication hubs within the community can streamline the flow of information and resources. These hubs, equipped with Baofeng radios and operated by trained volunteers or emergency response teams, serve as the heart of the communication network. They can manage incoming and outgoing communications, dispatch assistance, and relay important updates from local authorities or emergency services.

Integration with Existing Infrastructure:

Leveraging existing amateur radio infrastructure, such as repeaters, can significantly enhance the reach and reliability of the communication network. By programming Baofeng radios to access these repeaters, communities can extend their communication capabilities across the urban area and beyond, ensuring connectivity even when cellular networks are down.

Training and Education:

A critical component of establishing an effective urban communication infrastructure is training and education. Community workshops on the basics of radio operation, frequency management, and emergency communication protocols can empower residents to effectively use their Baofeng radios in times of need. Collaboration with local amateur radio clubs can provide valuable expertise and resources for these training programs.

Emergency Communication Drills:

Regular drills and simulations can test the resilience and effectiveness of the urban communication infrastructure. These exercises, ranging from simulated natural disasters to blackout scenarios, can identify weaknesses in the network provide practical experience to participants, and reinforce the importance of radio communication preparedness.

Frequency Congestion and Interference

Lastly, you need to know that the dense electronic environment of cities often leads to frequency congestion and interference. This can make it difficult for you to carry out radio communications effectively. But, like always, there are ways to tackle this problem. To use your Baofeng radio is such cases, focus on:

Frequency Planning and Coordination:

Developing a comprehensive frequency plan that assigns specific channels for different types of communication (e.g. emergency, logistical, community updates) can reduce congestion. Coordination with local amateur radio operators and emergency services to avoid overlapping frequencies is also crucial.

Use of Sub-audible Tones:

Implementing Continuous Tone-Coded Squelch System (CTCSS) and Digital Coded Squelch (DCS) tones can filter out unwanted transmissions and reduce interference. By setting up these sub-audible tones, Baofeng radios within the network can communicate more clearly, even in the presence of other signals on the same frequency.

Adaptive Communication Tactics:

In times of high congestion or interference, adapting communication tactics can help maintain network integrity. This can include using alternative frequencies, shifting to digital modes if supported, or employing relay methods where messages are passed through multiple radios to reach their final destination.

Interference Mitigation Techniques:

Identifying and mitigating sources of interference is a continuous challenge. Techniques such as repositioning antennas adjusting power levels, and utilizing directional antennas can help overcome interference from electronic devices, power lines, and other common urban sources.

Regular Network Maintenance and Monitoring:

Ongoing maintenance and monitoring of the communication network can preemptively address congestion and interference issues. Establishing a dedicated team to monitor network performance, resolve technical problems, and update frequency plans as needed ensures the network remains effective and responsive to the community's needs.

Conclusions

First things first, congratulations are in order because you've just completed the book and now you're an expert on Baofeng radios. Throughout the entire book we've learned quite a lot of things. Did you think the Baofeng radio would turn out to be such an amazing device?

We've gone on some adventures in this book and learned how the Baofeng radio can help us out in different circumstances. Be it emergencies, outdoor activities, being stranded on an island, or a few other things, the Baofeng is a gadget that'll always come in handy.

However, to make sure that you can use it effectively, you need to learn about its different features and functions. And that's exactly what we've done in the book. We've covered all the different communication modes on the radio, how you can set it to different frequencies and channels, and how you can customize it as per your own requirements.

But that's not all. We even learned how you can create a communication plan on your own and an antenna for different circumstances. But all of that was in the first half of the book. In the second half, we learned how you can use your Baofeng radio to be useful in emergency and survival situations and how you can master guerilla communications .

Your journey of learning about push buttons and radio channels and frequencies has now reached an end. As you close this book, remember that the skills and knowledge you've learned are not just for the crises that may never come.

They are for camping, hiking, trips to a remote island, and other outdoor adventures. As we say goodbye, I urge you to make the most of what you've learned. Take your Baofeng radio with you wherever you go and use it to communicate like the expert you are now.

And if you find someone who is developing an interest in Baofeng radios, share your knowledge and this book with them.

Good luck on your adventures and may your signals never fade away!

Made in United States
Troutdale, OR
10/01/2024

23309195R00069